T0203817

THE ECONOMICS OF SCIENCE: A CRITICAL REALIST OVERVIEW

Volume 1: Illustrations and philosophical preliminaries

Dramatic and controversial changes in the funding of science over the past few decades have stimulated a huge literature trying to set out an 'economics of science'. Whether broadly in favour or against these changes, the vast majority of these frameworks employ ahistorical analyses that cannot conceptualise, let alone address, the questions of 'why have these changes occurred?' and 'why now?'

This book argues that the fundamental underlying problem in all cases is the ontological shallowness of these theories, which can only be remedied by attention to ontological presuppositions. Accordingly, Tyfield sets out an introduction to the existing literature on the economics of science, together with novel discussion of the field from a critical realist perspective in ways that also develop critical realism as a philosophical project. The first of two volumes, this book explores substantive illustrations of the social challenges associated with the commercialisation of science before turning to a critical engagement with critical realism's critique of mainstream economics that underpins the project as a whole.

David Tyfield is a lecturer at the Centre for Mobilities Research and Sociology Department, Lancaster University. He is reviews editor of *Science as Culture* and formerly an editor of the *Journal of Critical Realism*.

Ontological Explorations

Titles in this series include:

THE ECONOMICS OF SCIENCE: A CRITICAL REALIST OVERVIEW

Volume 1:
Illustrations and philosophical preliminaries

David Tyfield

Routledge
Taylor & Francis Group

LONDON AND NEW YORK

First published 2012
by Routledge
2 Park Square, Milton Park, Abingdon, Oxon OX14 4RN

Simultaneously published in the USA and Canada
by Routledge
711 Third Avenue, New York, NY 10017

Routledge is an imprint of the Taylor & Francis Group, an informa business

British Library Cataloguing in Publication Data
A catalogue record for this book is available from the British Library

Library of Congress Cataloging in Publication Data
Tyfield, David.
    The economics of science: a critical realist overview/David Tyfield.
    p. cm.
    Includes bibliographical references.
    1. Science-Economic aspects. I. Title.
    Q175.5.T942011
    338.4′75–dc23   2011020419

ISBN: 978–0–415–49835–7 (hbk)
ISBN: 978–0–415–68879–6 (pbk)
ISBN: 978–0–203–16279–8 (ebk)

Typeset in Bembo and Stone Sans by
Florence Production Ltd, Stoodleigh, Devon

For Hannah,
without whom this book would
have been finished years ago,
and Adah, without whom it would
have taken even longer

CONTENTS

ILLUSTRATIONS

Figures

Tables

ABBREVIATIONS

ANT	actor-network theory
BIS	UK Government Department of Business, Innovation and Skills
BRICs	Brazil, Russia, India and China
CCP	Communist Party of China
CEO	chief executive officer
CMP	capitalist mode of production
CPERI	cultural political economy of research and innovation
CR	critical realism
EEI	evolutionary economics of innovation
ESK	economics of scientific knowledge
FDA	Food and Drug Administration
GCE	generalised commodity exchange
GM	genetically modified
HEFCE	Higher Education Funding Council for England
ICT	information and communication technology
IMF	International Monetary Fund
IMP	intellectual monopoly privilege
IP(R)	intellectual property (right)
IPO	initial public offering (of a company's shares)
KB(B)E	knowledge-based (bio-)economy
KLP	knowledge labour power
LTV	labour theory of value
MOST	Ministry of Science and Technology of China
NEGT	New Endogenous Growth Theory
NES	new economics of science
NIH	US National Institutes of Health

NKL	National Key Laboratory of China
NNPT	naturally necessary production time
NTTCs	National Technology Transfer Centres of China
OECD	Organization for Economic Cooperation and Development
OES	old economics of science
PhRMA/PMA	Pharmaceutical Researchers and Manufacturers of America
PLK	productive labour for capital
PRI	public research institution
rDNA	recombinant DNA
R&D	research and development
R&I	research and innovation
SCM	supply chain management
SEPM	synchronic emergent powers materialism
SNLT	socially necessary labour time
SNTT	socially necessary turnover time
SOE	state-owned enterprise
SSK	sociology of scientific knowledge
S&T	science and technology
STACOs	Offices for Commercializing Science and Technology Achievements
STEM	science, technology, engineering and maths
STS	science and technology studies
TA	transcendental argument
TNC	transnational corporation
TRIMs	Trade Related Investment Measures Agreement
TRIPs	Trade Related Intellectual Property Agreement
U-I (complex)	university-industry (complex)
UIRs	university-industry relations
UOTE	university-owned technology enterprise
URIs	universities and research institutions
USTR	United States Trade Representative
VC	venture capital
VTL	value theory of labour
WIPO	World Intellectual Property Organisation
WTO	World Trade Organization

PREFACE AND ACKNOWLEDGEMENTS

How should we respond to the increasing privatisation of scientific knowledge production and the academy more generally? What are the implications of this process, in fact? And why is this process occurring in the first place? How does it relate to other major changes and challenges in which scientific knowledge and technological innovation play a crucial role, such as responding to climate change or associated challenges of food security? It is in grappling with these questions over the past few years that I have come to formulate the arguments set out in this book. In the process, I have often felt the painful contradiction between the seeming urgency of these questions and the slowness and inefficiency of the research process. Indeed, this book has been many years in the making now. It started as a Ph.D. thesis, taking me back to the classroom after several years in the law, to investigate the implications of the World Trade Organization's extraordinarily iniquitous agreement on Trade-Related Intellectual Property Rights (TRIPs) for the funding of scientific research as a case study to test the methodological prescriptions of arguments within the philosophy of economics. Subsequent re-writing as a more general overview of the 'economics of science' then transformed the book into both a broader substantive exploration of the commercialisation of science and the contemporary neoliberal knowledge-based economy and a more concerted engagement with constructivist studies of science and technology (STS). The result, I hope, is a comprehensive argument for the importance of a theoretical synthesis of political economy and STS in which I do not just argue for, but also show, the gains from such concerted engagement.

Of course, none of this would have been possible without countless discussions with colleagues, only some of whom I can thank here personally. During my Ph.D., John Dupré, Steve Fleetwood, Francesco Guala, Clive Lawson, Tony Lawson, Uskali Mäki, Nuno Martins and Nigel Pleasants all offered very helpful comments on earlier drafts of various chapters, while I also had the benefit of fruitful discussions with Lorenzo Bernasconi, Jane Calvert, Adrian Haddock, Bob Jessop, Martin Kusch, Lenny Moss, Esther-Mirjam Sent and Matthias Varul. I would also like to thank Ros Webber for her assistance in numerous small but invaluable ways. More recently, in my happy and productive time at Lancaster, I have been privileged to have numerous discussions with John Urry, Brian Wynne, James Wilsdon, Robin Porter, James Keeley, Moxuan Li, Bob Jessop, Andrew Sayer, Ngai-Ling Sum, Bron Szerszynski, Larry Reynolds, Larry Busch, Bill Davies, Geoff Tansey, Bülent Diken, Kean Birch, Les Levidow, Phil Mirowski, Rebecca Lave, Ulrich Beck, Sheila Jasanoff, Sang-Hyun Kim, Adrian Ely and Andy Stirling. Finally, and most importantly, I would like to thank my parents, Stuart and Linda, whose support in countless ways has been so precious, and my beloved wife Deborah, for her patience and encouragement, without whom this work would not have been possible. I dedicate this book, however, to my beautiful daughters, Hannah and Adah. I hope in some small way it contributes to a brighter future.

SECTION I
Introduction

1

INTRODUCTION

Towards a critical political economy of science (or why the economics of science cannot be an economics of science!?)

1.1 Science funding in crisis

In October 2010, British scientific research received a hammer blow with the announcement of the Conservative–Liberal Democrat coalition government's *Comprehensive Spending Review*. With an unsustainably high fiscal deficit in the aftermath of the great financial crash of 2008 and the government bail outs of several British banks, including the rushed merger of what was, as a result, the world's largest, and the election of a government more attuned to austerity than Keynesian stimulus, a swathe of unprecedented cuts in the public purse of an average 20–25 per cent had been anxiously awaited through the summer, including at Britain's universities. On 22 October, the axe finally fell.

At the close of 2009, the (as-it-turned-out, outgoing) Labour government had already announced that it intended to cut public funding of science by £600 million by 2012–13; cuts that the University and Colleges Union had estimated would result in the closure of 30 universities and the loss of some 14,000 academic jobs (*The Economist* 2010a). Moreover, the Labour government commissioned a report, led by the former chief executive officer (CEO) of BP, Lord Browne, examining a 'sustainable future' for higher education. However, the early months of the new government ramped up the cutting rhetoric further, with Vince Cable, Secretary of State for the Department of Business, Innovation and Skills (BIS) that incorporates universities and science, suggesting the cross-departmental cuts of 20–25 per cent would also apply for the science budget. In the event, and following heavy and public lobbying by British scientists, the CSR announced a ring-fenced budget for the 'STEM' (science, technology, engineering and maths) subjects. This was met with little triumphalism on campuses, however, as the accompanying Browne Review, which was largely endorsed in the CSR, proposed an 80 per cent cut in core teaching for all other subjects and an unprecedented rise in undergraduate student fees from the current cap of just over £3,000 per year up to a maximum of £9,000. Demonstrations and protests have ensued in the following months, especially among current and future students outraged by the likely tripling in tuition fees. A particular target of this fury has been the Liberal Democrat partners of the coalition, many of whom, including the Deputy Prime Minister Nick Clegg, pledged to scrap tuition fees altogether in the May election.

While these events have taken place in a small, post-imperial country of around 62 million on the periphery of an 'old' continent, the UK is actually still a major player in global science, routinely running (albeit in aggregate, a distant) second place to the US as measured by scientific publications, citations and high-ranking universities. What is happening in British science is thus of more than parochial interest. But the squeeze on British (and especially English) universities is also not globally unique. In fact, just as the economic crisis has laid low economies across the developed global North (in particular), so too it has triggered a wave of consideration of how much public treasuries should really be investing in science (e.g. Chatham House 2010). These have not been as marked as the UK cuts, with notable exceptions; for instance, the state-funded University of California has suffered a 20 per cent cut in funding (approximately $1 billion), with student fees expected to increase by 20 per cent. Nevertheless, public funding of science is under the spotlight around the world, and will even suffer from the expiry of temporary post-crisis stimulus funds, as in the US.

The net effect of such reductions of public funding, however, need not necessarily be overall reductions. While these would certainly matter, the effect that is of more immediate interest to this study is the shift of science funding to private sources. This is, for instance, clearly the case with the recent reforms in the UK. As Stefan Collini (2010) notes, for all its importance and potential inequity, the rise in student fees is but a detail compared to the wholesale redefinition of science (both research and teaching) that the CSR reforms explicitly set out to achieve, namely the privatisation of British higher education (Finlayson 2010). By re-conceptualising higher education as the provider of business-relevant education ('skills') and research, hence the student and private research-funding source respectively as consumer and/or investor, these reforms augur the 'death of the university' (Couldry and McRobbie 2010), at least as we have known it to date.

Furthermore, while British scientific institutions, such as the august Royal Society (celebrating its 350th anniversary in 2010), have argued unequivocally for the need for generous public funding of science (e.g. Royal Society 2010) – and with some but limited success, as discussed above – their objections to these changes reveal only half-hearted opposition. For the core messages of the response to proposed cuts are phrased precisely to appeal to the same agenda, simply with conclusions inversed. Hence, the 'two urgent messages' of the Royal Society's public argument to this effect:

> The first is the need to place science and innovation at the heart of the UK's long-term strategy for economic growth. The second is the fierce competitive

challenge we face from countries which are investing at a scale and speed that we may struggle to match.

(Royal Society 2010: 4)

What is thus entirely absent is any discussion of the broader public importance of science, the need for public engagement in science and wider conceptions of higher education as 'education' as opposed merely to 'skills'. Similarly, regarding research, no objection is raised to another related, but quieter, revolution that is taking place in science funding in the UK.

In May 2009, with no great fanfare, the British Research Councils – the government-funded but quasi-autonomous bodies for competitive, peer-reviewed funding of science – introduced a requirement that all applications now include an 'impact statement'. In June of the same year, the Higher Education Funding Council for England (HEFCE) followed suit, announcing that the assessment of university departments for public funding would include an assessment of the 'impact' of the department's research. These seemingly vague, even anodyne, stipulations are unlikely candidates for an historic sea change. After all, what could be wrong with demands for accountability regarding the impact of research on the society that is sponsoring it? But by introducing an explicit condition that all projects have a clear goal of impact beyond the academy, with special focus on direct *commercial* and/or *policy* applications, and with a timeline as to when that will be realised, this extra criterion is likely to have far-reaching consequences.

In search of funding, researchers will face one of two equally unpalatable options. First, they can amend their research agendas to ensure that everything they are investigating can (attempt to) make credible and substantiated claims regarding short- to medium-term 'impact' of this sort. This could well undermine the development of theory that is relatively remote from practice, but which may offer considerable benefits to the development of knowledge (including the more practical work), as well as unpredictable and unquantifiable social effects, at a later date. Furthermore, in the social sciences especially, this thereby threatens to turn the profession from good researchers to 'crap consultants', as one senior policy academic frankly put it at the time. 'Where is the important theory on which I depend going to be developed now?' he opined.

The second option is equally corrosive, namely to fill in the impact statement as merely part of the 'game' of getting funding. At this point, the entire system of peer review – already suffering from problems (Henderson 2010), such as a bias of disciplinary panels against inter-disciplinary proposals (Nowotny *et al.* 2002), that often reduce success to simple luck regarding the allotted reviewers – risks becoming entirely debased, for no system built upon systematic falsehood, acknowledged by those within it, is sustainable, let alone likely to serve the purposes of its express design. For instance, the extra criterion will simply offer a further excuse for reviewers to strike down proposals that are strong in other respects but not to their theoretical liking, hence rendering the entire review process effectively arbitrary. Alternatively, competition regarding claims as to likely impacts could well lead to simple exaggeration. As our senior policy academic again puts it: 'Liars will succeed while good researchers offering reasonable assessment of their work will fail.' It is no surprise, therefore, that these proposals were also met with a barrage of criticism (if not media attention-grabbing protests, as with the rise in student fees) including a petition signed by over 16,000 academics, Nobel prizewinners among them.

1.2 The privatisation and commercialisation of science

These recent developments may be particularly striking and troubling, but they are by no means exceptions. Rather, they are simply the most recent examples of a much broader process of privatising and/or commercialising science (we will consider the distinction below) that has been proceeding for nearly 30 years now, in particular since a watershed of 1980. Furthermore, while this process has proceeded first and farthest in the US (hence a distinct bias towards the US in the literature (e.g. Lave *et al.* 2010)), it is effectively ubiquitous across the global North and increasingly prevalent even within developing countries, especially those with significant science bases, such as the emerging global powers of the BRIC countries (Brazil, Russia, India and China).

The increasing presence of business in science funding may be seen from two angles. First, we could explore the growing prevalence of private and/or commercial sources and mechanisms of funding within scientific research. Second, we could explore the growing emphasis placed upon a generic, abstract 'science' (or 'knowledge' or even 'information') within influential policy and business circles as the source of economic growth and competitiveness, via 'innovation'. 'Science' thus becomes increasingly legitimated and prioritised on predominantly (if never exclusively) economic grounds. In the context of capitalist market economies, however, this implies (or at least resonates with) subjection of scientific research to similar social mechanisms as for any other sector of the economy, hence the first perspective. Clearly, therefore, the second aspect of this process ties in closely with the prevalent and dominant policy discourses – that are also evoked, often entirely uncritically, in the social sciences – of the 'knowledge economy' and similar terms.[1]

There is little doubt regarding the evidence for these changes, which is ubiquitous and uncontroversial.[2] For instance:

Privatisation of research funding

Private funding of scientific research in the US has grown 3.8 times in real terms (8.7 times in nominal terms) since 1980 against increases of federal government funding of 1.5 times (and 3.5 times respectively). As a result, private funding of total research and development (R&D) has grown to 65–70 per cent of total national R&D expenditure in the decade from 1998–2008 from just under 50 per cent in 1980 (National Science Foundation 2010) (Figure 1.1), with most of this funding

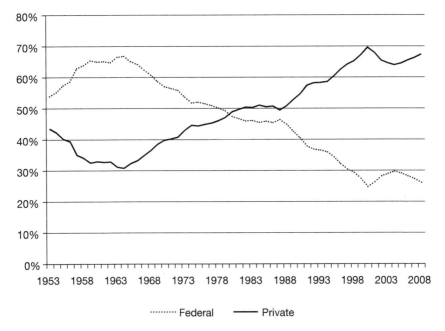

FIGURE 1.1 Total funding of US science by source, 1953–2008 (%)

Source: NSF (2010).

being directed to research itself conducted by private industry. Moreover, while growth in federal public funding of total R&D in the US was effectively negative throughout the late 1980s and 1990s, private funding grew strongly, especially between 1995–2000 when it grew at an average of 9 per cent per annum (see Figure 1.2). The picture is the same in the UK, where business funds 62 per cent of scientific research (£15.9 billion) in 2008 (before the public cuts), the remainder being the £7.8 billion of public funding for universities, £1.3 billion of public funding by government departments and £600 million from charities (*The Economist* 2010f: 44).

At universities, too, commercial funding has increased dramatically, with growth in private funding of academic research outpacing that of federal funding from the late 1960s and especially through the 1980s (average per annum growth in private funding of university science of 11.6 per cent between 1977–89) (see Figure 1.3). Accordingly, private funding rose to a peak of around 7.5 per cent of total university funding in the US in 1998–99 before dipping (with the bursting of the dot-com bubble and the aftermath of the events of 11 September 2001) and then climbing back to around 6 per cent in 2008. Global trends (notwithstanding significant national differences) also show a global average of around 6 per cent of university funding from private sources (Radder 2010: 9, quoting AWT 2005: 57) and a 200 per cent increase in external research contract funding between 1980–2001, excluding public research councils (AWT 2005: 45) against a backdrop of stalling public funding.

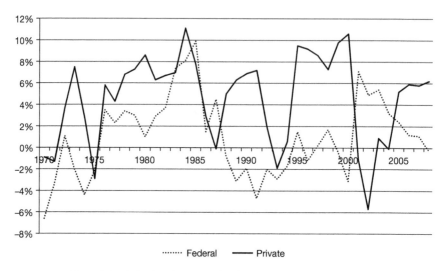

FIGURE 1.2 Growth in federal and private funding of total US R&D, 1968–2008 (% per annum)

Source: NSF (2010).

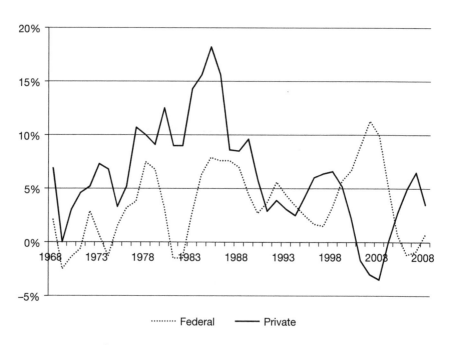

FIGURE 1.3 Growth in US university funding, 1968–2008 (% per annum)

Source: NSF (2010).

Commercial 'accountability' and 'relevance'/'impact' criteria in competitive public funding

The increased presence of private funding in scientific research, including in academia, is not the only way in which science has become increasingly subject to commercial pressures, commodifying science. Public funding, too, has increasingly come with strings attached that test the commercial relevance of research; an 'auditing culture' that seeks to quantify research achievements (Demeritt 2000, Shore 2008, Ward 2005). In the UK, this would include the Research Assessment Exercise of HEFCE, now rebranded as the Research Excellence Framework, that includes the new 'impact' criterion upon which senior government ministers have placed such emphasis in their reform of higher education funding. More generally, the transformation of the management of universities and public research institutions (PRI), with the emergence of ubiquitous management discourse associated with for-profit private enterprise, has shifted institutional imperatives within universities, hence conditioning the research and career goals of academic staff (Radder 2010, Kleinman and Vallas 2001).

Growth in university-industry relations (UIRs) and direct incorporation of science into commerce

Recent decades have also witnessed the proliferation of new and intensive connections between universities and private businesses, in multiple forms. At their most high-profile, and controversial, these take the form of privately funded centres or entire departments, perhaps as 'strategic alliances' contracted for extended periods of 5 to 10 years. For instance, the University of California at Berkeley has been the site of two major controversies centred on such proposals, namely a $500 million biofuels institute, funded by oil major BP (Blumenstyk 2007), and a $25 million biotech institute, funded by pharmaceutical giant Novartis (Rudy *et al.* 2007). Such alliances, however, hit the headlines precisely because they are not that common. Smaller projects and collaborations are increasingly ubiquitous, including the funding of individual Ph.D. projects. In most such cases, as for the larger institutional tie-ins, funding will come with the funder's ownership of the research results and all this entails, regarding rights of oversight of research publications, first rights to patent and license findings and direction of future research that uses the results. Finally, and especially in fields relating to hi-tech industries such as biotechnology or information technology, universities have encouraged their staff to build on research findings to spin-off start-up firms. The result, especially at the most successful research universities, has been a revolving door between university and commerce, as senior faculty take on dual roles as university research leaders and entrepreneurs or company officers.

Growth in patenting, especially at universities and especially in life sciences

In the post-war/cold war period, intellectual property rights (IPRs) played a comparatively limited role in the everyday practice of most university scientists, who instead aimed at public dissemination of their results via peer-reviewed publication. Since the early 1970s, however (and thus since *before* the passing of the US Bayh-Dole Act allowing patenting of publicly funded research (Mowery *et al.* 2004)), patenting at universities, especially in the US, has grown rapidly. Moreover, this growth has been particularly marked in high-growth sectors of science-intensive high technology, such as biotechnology.

Commodification of higher education

Science education has also become progressively privatised, with student fees an increasingly important source of revenue. This has led to the transformation of students into 'consumers' of higher education, with commensurate focus on improving the 'student experience', especially through the construction and provision of high-quality catering, accommodation and leisure facilities, and the concerted marketing of universities as global brands, especially through intra-university sports and sportswear (Slaughter and Rhoades 2004, Jessop *et al.* 2008). In the US, in particular, there has also been a significant growth in for-profit higher education, rising some 59 per cent in 3 years to 3.2 million students at 3,000 colleges in 2008–9, and representing 11.7 per cent of all American students (*The Economist* 2010d: 13), making for-profit higher education 'one of the greatest success stories in American business' (*The Economist* 2010e: 63).

Strong, global intellectual property rights, especially for life sciences and ICT

Last, but by no means least, is a development that is seemingly distant from the commercialisation of science but, as we shall see, is of immense significance (see Chapter 4). In 1995, the founding of the World Trade Organization witnessed the unprecedented introduction of international law for strong IPRs, despite vociferous opposition from almost all developing countries, particularly India and Brazil.[3] This 'Trade-Related Aspects of Intellectual Property' agreement (abbreviated TRIPs) was an extraordinary development on many levels, if not a veritable commercial coup (see e.g. Sell 2003, Drahos and Braithwaite 2002, Dutfield 2003, Richards 2002). It instituted strong IPRs that even pro-TRIPs economic analyses have shown would benefit only a handful of (largely US-based) transnational corporations, particularly those in the IP-sensitive industries of pharmaceuticals, agribusiness, information and communication technology (ICT) and entertainment (e.g. Maskus 2000). Enacting such strong IPRs on developing countries also trumped historical precedent in which nation states have tended gradually to strengthen IP law to reflect the strength of domestic industries dependent upon them (Chang 2002, May and Sell 2006). Second, it was drafted and orchestrated by a tiny coterie from these businesses, despite the fact that they were not even parties to the agreement, not being sovereign nation states (Sell 2003). Finally, it made explicit provision to compel the introduction of IPRs for the life sciences, including pharmaceuticals, for which patents are *most* controversial regarding their effects on public health systems.

In fact, there is no clear *ex ante* or philosophical reason why the increasing prevalence of scientific research done within or funded by private industry should be seen as a problematic phenomenon (Shapin 2008). For instance, it is increasingly difficult to maintain any neat correlation between the institutional location of scientific research and the kind or standard of research; for example, (publicly-funded) universities doing cutting-edge theoretical or 'basic' research while private R&D labs do less advanced applied tinkering, with Nobel prizewinning science conducted in private laboratories and university teams working on 'applied' sciences (see Section IV in Volume 2). Nevertheless, together all these changes have raised serious concerns about the future of scientific research and its institutions (especially the university). For instance, Radder (2010: 14) lists eight issues that have attracted critical comment:

1   The potentially undesirable influence of commercial interests on research methods and results.
2   Higher levels of secrecy as scientific findings are transformed into commercial secrets. For example, there have been a number of high-profile cases of academics falling foul of their institution's commercial sponsors at University of Toronto, University of California San Francisco and Memorial Hospital of Brown University (Krimsky 2003).
3   Downgrading of research disciplines not seen as relevant from the perspective of profitable economic activity.
4   A short-termism in research agendas, as commercial investment demands quick pay-off, to the detriment of longer-term 'basic' research or other socially beneficial projects demanding longer-term horizons.
5   Assorted objections (ethical, legal, philosophical, religious, etc.) to the patent-ability of academic research, especially those associated with the life sciences, which may raise questions about the integrity, sanctity, dignity and/or self-ownership of organisms, especially humans.
6   Conflicts of interest and exploitation of public funds for private gain by entrepreneurial scientists.
7   Detrimental effects on public trust in science more generally and the (seemingly) 'disinterested' epistemic authority of scientific findings.
8   General concerns regarding the 'justifiability of the privatisation and economic instrumentalisation of public knowledge'.

To this we may also add, regarding for-profit higher education, criticism regarding the tendency of such institutions to focus on profit maximisation to the

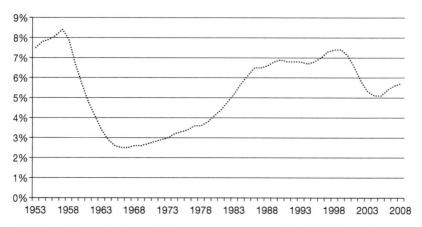

FIGURE 1.4 Private funding of US university research, 1953–2008 (% total)
Source: NSF (2010).

detriment of learning, hence deploying a shoddy, rapid-turnaround and high-volume model of commoditised course materials – 'digital diploma mills' (Noble 2002) – that also employs a cheap, temporary and exploited academic workforce. Indeed, recent criticism of for-profit higher education in the US, where it is most advanced, has reached a fever pitch, with one hedge fund manager telling a Congressional hearing, 'Until recently, I thought that there would never again be an opportunity to be involved with an industry as socially destructive as the sub-prime mortgage industry. I was wrong. The for-profit education industry has proven equal to the task' (Steve Eisman, quoted in *The Economist* 2010d: 13).

Moreover, as Kleinman (2003, 2010) stresses, these impacts need not be the effect only of direct intervention of private investment and finance in previously publicly funded research institutions, but may also, if not *primarily*, arise from a more pervasive and indirect transformation of academic research cultures. After all, at only 7 per cent of university funding (or even at 25 per cent in the fields where it is most concentrated), private funding is by no means dominant to date or even historically unprecedented (see Figure 1.4). Yet, these figures alone do not adequately capture the extent to which, in response to the increasing economic dominance within total science funding of private enterprise funding science that is itself done by business, both (i) researchers have themselves adjusted their research agendas and practices (e.g. patenting) in ways that would make their work attractive and acceptable to potential private funders *regardless* of whether or not it is so funded (Kleinman 2010, Welsh and Glenna 2006), especially in the context of reforms that have undermined the financial security of university researchers and the career and institutional demands to secure external funding (Collinson 2004, Nixon *et al.* 2001); and (ii) even public funding has been qualitatively transformed in its priorities and demands in the same direction (as discussed above).

From this perspective, however, it is arguable that the problems attending the commercialisation of science are even greater than those listed above associated with the straightforward growth of private funding per se. For it is possible that the continuing commercialisation of academic science may undermine the institutional integrity of the university (Brown 2010, Fuller 2010), the existence of a socially distinct institution for the production of diverse, original and critical knowledge,[4] and thus even, paradoxically, the social foundations of the productivity of knowledge demanded by contemporary capitalist knowledge economies.

1.3 Understanding the changing economics of science . . . or not . . .

In short, therefore, there have been exceptionally broad and far-reaching changes in the economics of science (and on a global scale) that have generated a pervading sense of crisis. It is no surprise, therefore, that in response to these concerns and the apparent rise of the 'knowledge economy' there has been a proliferation of projects devoted to understanding the 'economics of science'. Two overlapping literatures are particularly clear: first, the interdisciplinary debate regarding the commercialisation of science per se; and second, a series of broader, less polemical projects under the title of an 'economics of science'. We will return to the former shortly. Yet, as regards the latter, there are almost as many definitions of the 'economics of science' as there are practitioners (e.g. Sent 1999). Certainly, there is no clear emerging paradigm of an economics of science that is yielding important insights and attracting a growing momentum of research findings. Furthermore, none of the projects that *have* received widespread attention are concerned with exploring and explaining the changing economic underpinning of scientific research as a historical process with profound social repercussions. In other words, the crucial questions of 'why *these* changes in the economics of science, in *these* places and *now*?' are almost entirely elided by such work. In the absence of such explanatory (and possibly critical) work, however, it seems hard indeed to attempt to forecast possible effects of these developments, whether normatively negative, neutral or positive. As Sent (1999: 112) has claimed, therefore, 'many construals of an economics of science are inadequate to the task of understanding the dynamics of science funding', including those that have dominated the literature to date.

What, then, do these high-profile projects focus on? Among the range of 'economics of science' that have received most attention, two approaches in particular stand out. First, there are various projects that employ mainstream economic analysis to investigate the institutional conditions for the optimal allocation of resources in order to maximise output of scientific research or scientific productivity.[5] Second, a number of projects within philosophy of science have sought to use economic analysis, particularly neoclassical economics, to resolve perennial problems of epistemology regarding the need to square the possibility of 'truthful' knowledge with the social particularity of its inception. These projects aim to 'naturalise' philosophical debate by grounding it in (social) scientific methods that provide *scientific* justification for *philosophical* conclusions.[6] In this case, the 'economics of science' thus deploys economic models of science involving rational, utility-maximising individuals concerned to maximise their 'cognitive' resources; what has been called an 'economic epistemology' (Mäki 2004). According to proponents of this position, these models thus provide clear explanation of how self-serving individual scientists can nevertheless produce socially advantageous 'true' scientific knowledge, instead of succumbing to the temptations of short-term riches through scientific fraud, plagiarism, etc.; in sum, an epistemic 'invisible hand' argument akin to that regarding the supposed emergence of public virtue from private greed as in standard market models of the economy since Adam Smith.

Despite their differences, we may note straight away that both projects are 'economics' not in the sense of exploring economic *aspects* of science but rather in employing forms of economic *analysis* (Sent 1999).[7] They thus have nothing to say about the fundamental questions regarding the funding of science, for this is not their primary goal. As such, it may be tempting simply to bypass this work altogether and proceed directly to alternative projects, perhaps not explicitly titled the 'economics of science', that have much more to say about these concerns. Indeed, we will not dwell in any detail on the latter group of philosophical projects, not just because their project is so loosely related to our concerns here regarding explanation of science funding but also because their enterprise is highly problematic on its own terms.[8]

But we cannot pass so lightly over the former group, for it raises the crucial question 'where *is* the economics of science?' Or rather 'why has (mainstream) economics failed to develop an economics of science?' Consideration of this point shows clearly that an economics of science poses a serious test to the entire conception of economics underlying the mainstream of this discipline. As a result, we may not only definitively repudiate mainstream economics as a profitable (!) avenue of enquiry for an economics of science, but we are also thereby thrust into the philosophical (ontological, epistemological and methodological) questions of what should replace it.

1.4 An 'economics' of science?

Let us, therefore, consider a project within mainstream economics that is widely seen as a seminal reference point for the development of an economics of science, namely the 'new economics of science' (NES) of Partha Dasgupta and Paul David (Dasgupta and David 1987, David and Dasgupta 1994). Not only are these two highly esteemed economists, now semi-retired emeritus professors at Cambridge and Oxford respectively – the former earning his spurs in work on the economics of information with Nobel prizewinner Joseph Stiglitz and the latter also a central figure within the broader 'Stanford' school of economics of technological change that has been highly influential and demonstrably successful in developing understanding of themes closely related to an economics of science – but their 1994 paper is a singularly highly referenced work seen as the founding of a new approach to these issues. It therefore serves as a representative of a broader programme of work that includes other high-profile work such as Dominique

Foray's 'economics of knowledge' (2004). Finally, as an explicitly *post*-neoclassical project, their NES work represents definite progress upon jejune neoclassical attempts and is not amenable to many of the criticisms that can be levelled against that framework.

Dasgupta and David's economics of science is 'new' in that it builds upon and replaces an 'old economics of science' (OES). The latter discourse is effectively captured in a handful of extremely highly cited papers from around the early 1960s (Arrow 1962a, 1962b, Nelson 1959) that used neoclassical economic analysis to argue for (generous) state funding of scientific research. This argument was based upon the identification of a 'market failure' in the private funding of (the production of) scientific knowledge. Against a presumption of market-based economic organisation, therefore, an exceptional case can be made for the need for state funding.

Arrow explains that, as knowledge is a 'public good' whose appropriability is limited, there will necessarily be a shortfall in its private production from the optimal level due to the absence of sufficient incentive for individuals to fund it. A public good, so defined, has several characteristics, namely 'nonrival possession [i.e. one person can own or use it without depriving another person], low marginal cost of reproduction and distribution (which makes it difficult to exclude others from access), and substantial fixed costs of original production [so that there is significant disincentive to produce it without expectation of commensurately high return]' (David 1993: 27). Together, these characteristics undermine any incentive for provision by private enterprise. Accordingly, there will be market failure in the funding of science and technology research. Yet, the market failure is even greater in the case of markets for information than for material public goods such as traffic lights, because information/knowledge has the following special characteristics:[9]

- Extreme indivisibility and durability, so that once discovered or received, 'there is no value to acquiring it a second time' nor any 'societal need to repeat the same discovery or invention because a piece of information can be used again and again without exhausting it' (ibid.: 25).
- Asymmetry in information regarding the value of the knowledge, in that 'the attributes of the commodity – typically, the complete context of the information itself – will not be known beforehand' (ibid.: 28). This adds considerable complication to the negotiation of contracts for information and, hence, exacerbates market failure.
- 'Its cumulative and interactive nature' (ibid.), by which the private parcelling of knowledge/information proscribes the possibility of productive interaction between disparate ideas in a way that does not occur in separate treatment of municipal parks or traffic lights.

From this analysis, it follows that adequate levels of science funding will only be attained if provided by the state, as embodiment of the public (as opposed to individual, private) interest. Yet, even if science is a 'public good', why is it a priority for government spending? To complete the old economics of science, therefore, this analysis was matched with an argument that 'basic' science seeds the tech-nological innovations that, in turn, drive economic growth; a 'linear model' of economic growth with science at its source. A US science policy document pub-lished at the end of the Second World War entitled *Science: The Endless Frontier* by Vannevar Bush (1945) is usually cited as the seminal text for this part of the argument.[10]

This orthodoxy regarding science funding remained effectively unchallenged for several decades in the post-war period, matching the economic reality of generous state funding of science, particularly in the US. Indeed, the literature on the economics of science during this period is so brief not least because it was widely believed that there was little else to add. Instead, such research as there was in related fields within economics focused on the effects of different (or 'asymmetric') information between parties involved in market exchanges on the vaunted capacity of markets to reach optimal allocations of resources, i.e. on further market failures related to the information of market participants (as in Dasgupta's earlier work).

Yet, the profound changes in science funding that began to erupt from the late 1970s provided the motivation to re-open the economics of science. As political arguments and economic reality were increasingly shifting science funding away from the public sector to private enterprise, the role of these two forms came under renewed scrutiny. Furthermore, developments within related disciplines had begun to undermine both pillars of the argument. As regards the market failure argument, the 'public good' aspects of knowledge were increasingly being prob-lematised: was it really true that knowledge could not be appropriated and so would not be optimally produced by private enterprise? Such doubts were related to criticisms of the central assumption of the entire OES framework that knowledge could be treated just like any other 'commodity' (hence the interchangeability of 'knowledge', 'information' and 'science'). In particular, the distinction between codifiable, and hence publicly available, and tacit, embodied knowledge began to receive much greater attention (e.g. Collins 1974, Johnson *et al.* 2002, Nonaka 1994, Pavitt 1998, Polanyi 1969), leading to a stronger distinction between infor-mation and knowledge respectively. While the OES characterisation of knowledge as non-rival and non-appropriable seemed to fit the former case, the embodied knowledge and skills of the scientist or expert seemed to be very difficult to transfer but no less crucial for scientific and technological advance. Moreover, even codified information generally requires knowledge capacities before it can be used effectively. The market failure argument regarding knowledge production thus seemed to be significantly weakened.

Furthermore, regarding the linear model, research from economic history, the history of technology and innovation studies (a discipline in which post-neoclassical, evolutionary economics has been pivotal) was repeatedly demonstrating that, even if economic growth depended upon technological innovations, 'basic' scientific research was rarely the source of these developments. Rather, most such techno-logical innovation occurred at its own level and only in a few industries, particularly pharmaceuticals, was direct interaction with scientific research at all significant (e.g. Levin *et al.* 1987, Mansfield 1986, Rosenberg 1974). These findings, in turn, resonated with developments within the philosophy and sociology of science that were emphasising the difficulty, if not impossibility, of providing an 'internalist' definition or demarcation of 'science', i.e. according to intrinsic characteristics regarding the kind of knowledge it produced and its epistemically superior 'scientific method', as had been presumed by the positivism that had dominated post-war philosophy of science.[11] From this perspective, therefore, science should rather be defined according to 'externalist' criteria of its social context and mechanisms of organisation, and scientific knowledge was no longer seen as privileged above technological knowledge. The OES framework was thus becoming increasingly untenable without significant revision.

It was in this context that David and Dasgupta began to explore an alternative theoretical framework for the economics of science. Their 'new' economics of science thus effects two major changes to the old version reflecting the developments just described. On the one hand, it incorporates a broader definition of 'knowledge' that admits its 'tacit' dimensions. On the other, science is defined sociologically, according to the social norms that set it in contradistinction to the 'other' form of knowledge, technology, which is treated with parity regarding both its epistemic status and socio-economic importance. Hence 'Science' (the term is capitalised) is defined by norms, originally identified in the sociology of science of Robert Merton (1973), such as rapid and open disclosure of research results, while 'Technology' (ditto) is that knowledge work that is dominated by the goal of building profitable commodities and so is marked by private appropriation and non-disclosure. Science and Technology (S&T) are then treated as a functionalist pair, in which both interact in myriad complicated (and unspecified) ways, and both logics are needed in order to maximise their respective contributions to economic growth.

This 'new economics of science' then aims to provide economic models – employing the latest developments in game theory and other branches of economic inquiry regarding the interaction of self-seeking, rational maximisers with limited information and/or cognitive capacities (or 'satisficers') (e.g. Simon 1979) – that explain the emergence, stability and problems of the distinctive norms of Science. Two goals may, thus, be served by this investigation, it is claimed. At the macro level, the economic structures that afford a productive scientific sphere may be identified and hence secured by policy, while at the micro level, more detailed models may be used to explore particular problems regarding inefficiencies of productivity and their potential policy solutions. The former dominates, however, with the overall conclusion being advice to proceed with caution regarding the commercialisation of science lest it transform too much activity within Science into Technology, thus defined, and hence undermine the functionalist balance between these two with implicitly disastrous effects for them both.

This NES certainly represents an improvement upon the old version regarding its acknowledgement of the problematic assumption of the identity of knowledge and information. Yet, it remains highly problematic in many respects, of which we list here only a few of the most significant. First, while a major element of its criticisms of the OES is the inadequate definition of knowledge employed, in fact the NES does little to break with that framework. Indeed, even while acknowledging the importance of tacit knowledge, the NES still treats 'knowledge' in its models as an uncomplicated commodity that is to be maximised (Mirowski and Sent 2002b; see Chapter 2). Qualitative issues regarding the effect of funding upon *which* avenues of scientific enquiry are pursued and the even more serious possibility that the very character of 'knowledge' itself may be dramatically debased by certain funding arrangements (consider, for instance, the possible reduction of 'science' to partisan corporate advertising) is simply overlooked (Mirowski 2011).

Second, while a welcome step in the right direction, the sociology of science employed in this project invokes a model of this discipline that is now highly controversial, if not definitively repudiated by the contemporary mainstream.[12] In short, the Mertonian sociology of science saw its task as the identification and explanation of the distinctive norms of science that afford the context for the pure pursuit of the scientific method, the latter being identified by the separate discipline of the philosophy of science. Conversely, a generation of work in science and technology studies (STS), beginning with the sociology of scientific knowledge (SSK) from the early 1970s, has shown how social factors penetrate right to the very heart of scientific research, and indeed, are inseparable from it (see Chapter 9). 'Social influences' on science are not, in short, only relevant when science goes wrong or is corrupted in some way and the entire hypothesis of defining science by its distinctive 'norms' has largely (and justifiably) been abandoned.

Third, and most importantly for our concerns, one may also criticise the project as a whole. It is clear that the fundamental purpose of the NES is exactly the same as that of the old version, namely to explore questions of allocative efficiency for policy in order to maximise scientific output. Neither project therefore aims to develop understanding of the actual economic processes of science funding. Furthermore, employing an ahistorical and rationalistic methodology, neither is capable of even conceiving, let alone asking, questions regarding explanations of *changes* in patterns of science funding. On the one hand, if actual funding arrangements match the prescriptions of these frameworks, they are not only 'explained' but also thereby justified. Conversely, where the actual funding of science does not fit, it is both unjustified and effectively inexplicable. As such, these frameworks are entirely incapable of explaining changes in actual funding mechanisms, except in the most anodyne Whiggish sense, for they are incapable of accounting for the period in which these diverged more significantly from the theoretical prescriptions: how are they as they are now or how were they ever otherwise? Similarly, where they are critical of changes, their analysis is largely impotent given its lack of explanatory power.

Finally, and perhaps most damning from its own perspective, is the fact that, despite some 15 years of intense research activity regarding related issues, there remains little apart from the original papers to show for the development of a broader economics of science programme. One may have expected that, were this a genuinely fruitful line of enquiry, there would by now be clear demonstrations of its analytical strengths, not least because it has been so highly cited. Yet, a mainstream (even post-neoclassical) economics of science remains curiously but undeniably elusive. As Mirowski (2009) puts it, 'the landscape [of the mainstream economics of information], far from being crowded with monumental theorems and general models, is merely dotted with abandoned half-finished shells'.

Indeed, exploring why this is the case indicates even bigger problems with such a framework that seem to render it a dead end. These problems hinge on the intrinsic difficulty, if not impossibility, of exploring the economics of science (or knowledge or information) using mainstream economic models built upon market exchange. Yet, illustrating the difficulty, if not impossibility, of perfectly functioning markets of knowledge also has significant implications for the very process of the actual commercialisation of science that an effective, explanatory economics of science would (currently) be examining. For it is towards just such an economy of science that these processes have been developing, under the quintessentially neoliberal banner of the 'marketplace of ideas' (Mirowski 2011). Indeed, the actual process of the marketisation of science has traded on the simplistic appeal of such markets, even as economic research has continually stumbled over the problems and paradoxes such markets present and, indeed, sometimes expressly sought to challenge these developments.

At its strongest, this line of criticism runs as follows (Boyle 1996). The most simple market model investigating the equilibrium of supply and demand for a commodity is built upon the assumption that individual agents have perfect information. As the access to information becomes more limited, the models become more complicated and less mathematically tractable, while such supposed mathematical rigour and analytical parsimony is a primary appeal of these models. Yet, if the commodity being traded is itself information, the market model must break down altogether, for it is impaled on the horns of a dilemma: either information held by each individual is not perfect (or some approximation of this), so that market exchange towards equilibrium cannot take place at all because no one is prepared to trade – a willingness to buy or sell can only be evidence that the buyer/seller knows something you don't; or information is perfect, in which case why would there be a market to purchase what one already has?

Even if one is sceptical of such a conclusive *coup de grâce*, swiftly putting an invalid 'economics of science' out of its misery,[13] there are several considerations that greatly problematise the possibility of such a mainstream economic project, even while interest in an economics relevant to the 'knowledge economy' grows ever more intense. Mirowski (2009), for instance, lists four such problems (in characteristically vivid prose), namely 'the impossibility of having your cake and eating it', 'the curse of the schizophrenic agent', 'the Wizard of Oz effect' and 'the broken bootstrap'. Each of these proposes theoretical objections to the possibility of an economics of information using such forms of economic analysis, on the basis of internal contradictions that arise from any such attempt. Regarding 'the Wizard of Oz effect', for instance, he notes (ibid.: 138–139):

> You cannot paint the marketplace of ideas as a marvellously parsimonious and magnificently efficient model of cognition if you can't even demonstrate mathematically that the internal production of neoclassical market equilibrium does not bear the information requirements that outstrip any other known algorithmic process.

Similarly, regarding 'the broken bootstrap', 'the price of the marketplace of ideas leads to formally undecidable market prices'.

In short, there are good reasons, reflected in the actual, historical experience of the various projects attempting a (post-)neoclassical-based economics of information, to suppose it is an impossible task, with significant implications also for the attempted construction in reality of such a 'marketplace of ideas'. Certainly, as Mirowski (2009) himself notes, such principled, even philosophical, objections to such a project are unlikely in the extreme to alter the work of most mainstream economists, not least because of their widespread disdain for philosophical and/or methodological discussion. Given these antinomies, however, those of us who are not indissolubly committed to market-based analysis may choose instead to seek for an economics of science from more propitious starting points. Moreover, an ongoing dissatisfaction with the neoclassical orthodoxy itself has resurfaced as something of a crisis in recent months within economics, among both faculty and students, in the context of the inadequacies of the mainstream discipline to deal with and explain the current global economic crisis (e.g. Fullbrook 2003, Harcourt 2010). The drive towards an alternative economics of science is thus indisputable.

1.5 An economics of 'science'?

We may, therefore, accept the challenge of searching for an alternative economics of science based on questions regarding the aims and tractability of the economics involved. But similar conclusions may also be reached by consideration of the second half of this phrase, namely 'what is the economics of science an economics *of*?' We have already briefly seen the problems raised by assuming a mainstream economic perspective regarding this issue. For taking such a stance, and hence seeking to understand 'science' in terms of a market, necessarily demands that there be some 'thing' that is produced by science and that it is self-evidently a social good to maximise. From this starting point, it is extremely hard *not* to proffer models that reduce science to a familiar commodity, which may interchangeably be called 'knowledge' or 'information', at least not without bringing the usefulness of this approach fundamentally into question.

Yet, this entirely fallacious reductionism also comes at a large analytical cost that significantly undermines the value of any conclusions. This would include any policy recommendations, which would likely produce major and serious unintended effects if implemented, given the huge divergence between the highly idealised assumptions of such models and the phenomena they are purporting to explain. For instance, a purely quantitative definition of the output of science not only significantly distorts and underestimates the varied and often entirely unpredictable contributions of such research to the advancement of knowledge and/or public welfare but also completely overlooks crucial qualitative and normative questions regarding the directions of research agendas and the balance of funding between (potentially competing) lines of enquiry. Similarly, in its analytical demand for a single, uniform and well-defined 'science' it discounts the significant differences across sciences, thereby necessarily excluding some sciences from consideration by simple definitional fiat. Furthermore, as discussed above, it entirely elides the crucial question regarding the effects of different funding arrangements upon the very nature of the scientific enterprise and the knowledge thereby produced. This effectively precludes an entire programme of research, which, if we would like to distinguish between a focus on scientific and economic developments, may be called an 'economics of *scientific knowledge*' (ESK) as opposed to an '*economics* of (the institution(s) of) science'.

Unsurprisingly, perhaps, these considerations are particularly apparent when we shift slightly from consideration of self-styled 'economics of science' projects to the second literature mentioned above, namely the closely related debate about the current privatisation of science. For in its direct concern with science, this literature is intrinsically more attuned to concrete concerns regarding how and why science is funded. It also raises to the fore the crucial questions regarding the impact of changing economic arrangements on the science itself, and so poses more directly the problems that arise from the extremely 'thin' definition of 'science' employed by a mainstream economics of science and its uselessness in studying actual sciences and their economic organisation.

Unfortunately, however, much of the literature on the commercialisation of science is just as problematic regarding its perspective on the nature of science and the interaction of 'science' and 'money'. Mirowski and Van Horn (2005) describe this literature in terms of a debate between 'Economic Whigs' and 'Mertonian Tories'. The former, in fact, like the mainstream economics of science (i.e. the economics of science employing mainstream economics) employ economic analysis that is entirely incapable of exploring the impact of funding arrangements on scientific knowledge. Indeed, given the need of such analysis for an *ex ante* definition of knowledge as a fixed and uniform thing ('knowledge' or 'information'), this approach actually rules out investigation into any *changes* of science that arise from changing economic arrangements. Instead, such Economic Whigs are simply concerned with maximising the productivity of 'science' and, true to their Whiggish economic (neo-)liberalism, tend to promote the commercialisation of science as a progressive development without any complications or problems for scientific research.[14]

Conversely, the latter adopt the Mertonian perspective of science being dependent upon social norms that leave it in splendid isolation from the corrosive influences of commerce and self-interest.[15] The commercialisation of science is thus treated as the catastrophic passing of a former golden age (i.e. the post-war period of the *trente glorieuses* of 1945–75) in which the state 'wisely' chose to fund science generously for the public good. Although this approach shows a much clearer concern regarding the interaction of changing economic arrangements and the scientific enterprise, it too is highly problematic. For instance, in its Mertonian assumptions that public funding preserves science as a pure realm of disinterested enquiry, it completely overlooks the strong military rationale, arising from the cold war, for much of the public largesse of its vaunted golden age (Dennis 1987, 2004, Krige 2008, Mirowski and Sent 2008). The history of science and more recent developments in the social studies of science and technology have also conclusively demolished the notion of 'science' it employs, systematically excluding socio-political concerns except insofar as they are distortions or corruptions of the scientific enterprise. In short, science has *never* been such a pristine 'sphere' of human activity and, indeed, such a conception is arguably unintelligible (see Chapter 9). Moreover, this form of analysis also mobilises an implausibly uncritical and undifferentiated conception of the state, while also ignoring the need for a political rationale for public funding; as the present fiscal crisis makes painfully clear, public funding of science must deploy money from *somewhere* and legitimised in *some* way.

The Mertonian perspective is thus merely the flipside of the errors of the Economic Whigs, both frameworks effectively ruling out the investigation of the actual effects of different funding arrangements on science, if for diametrically opposed reasons. In both cases, this arises from their respective definitions of science as an unchanging thing: for the latter, the thin and fixed conception of science as the 'knowledge' output (as we have seen); for the former, an inviolate social sphere that is either pristine or bust. Hence, while Economic Whigs cannot conceptualise any possible concern regarding the impact of economics on science, the opposite is the case for Mertonian Tories: 'state' (i.e. public funding) is good and 'market' is necessarily bad as irredeemably inconsistent with the given and fixed norms of 'Science'. In both cases, therefore, there is no need for (let alone possibility of) empirical investigation into the actual effects, both negative *and positive*, on science of changing economic arrangements because the answer is already known.

It is clear, therefore, that if we are interested in actually investigating questions such as 'how is/are science(s) funded?', 'how and why has this changed?' and 'how have these changes affected that/those science(s)?' we must employ a completely different conception of science. Nor does the solution to this problem lie in finding some middle way or compromise between the two perspectives. After all, this is effectively what David and Dasgupta attempted with their 'new economics of science', employing economic analysis to justify their Mertonian scepticism, and we have seen how this project fails. Rather, the source of the problems of these two frameworks must be tackled head-on, namely the presumption of both frameworks regarding the 'thingness' of science, which ensures that science is treated as an immaculate and *sui generis* entity. For as soon as this step has been made, the interaction of science with socio-political forces must necessarily be as a supererogatory and probably corruptive influence, and so detailed investigation into their actual interaction is forestalled.

Conversely, by directly challenging this fundamental presumption, an alternative conception of science can be elaborated that altogether bypasses the problems of both sides of the familiar dualism, and does indeed afford analysis of the relevant questions. This perspective would not only recognise the variety of social practices designated 'science' and attend to their concrete particularities, but it would also acknowledge that science is itself *constituted* as an irreducibly socio-historical process, with all the economic, cultural and political 'thickness' this entails. As we shall discuss in detail in Section IV in Volume 2, this vision of science is closely akin to that developed with STS under the rubric of 'co-production' of science and 'society', i.e. the mutual constitution of relatively autonomous social phenomena (e.g. Jasanoff 2004).

Such a redefinition of 'science', however, also brings with it significant consequences for the form of economics that is capable of studying it. As Table 1.1 summarises (albeit simplistically, for the sake of didactic expositional clarity), there are clear connections between the conception of 'science' (as research object) employed by an economics of science and its conception of the 'economics' studying it. Yet, this relation works in both directions. Changing the definition of science thus, of necessity, imposes conditions on the economics relevant to study it.

Three points, in particular, illustrate this relation and the need to move towards the bottom of the table. First, on top of all the objections we have been considering so far regarding the top two rows, we may also note that it is only in the bottom row that the two columns actually overlap and so come into contact. Only when taking this perspective, therefore, is there an 'economics of science', i.e. an economic *aspect* of science, that merits and affords investigation.

TABLE 1.1 Relating 'economics' and 'science'

Framework	*'Economics'*	*'Science'*
Economic Whig	Neoclassical economics, maximisation by rational agents	Thin conception of science as thing regarding output as 'Knowledge' or 'Information'
Mertonian Tory	Social institutions relating to profit maximisation (e.g. NES's 'Technology')	Science as social thing: social institution relating to maximum disclosure of 'knowledge'
Explanatory, critical economics of science	Economics as study of political economic forces within society more broadly	Science as socio-political process of 'knowledge' production, circulation and consumption

Second, once science is conceptualised in terms of co-production, it becomes apparent that not only is the trajectory of scientific development affected by its socio-economic and political context (itself developing in interactive parallel), hence affording exactly the kind of analysis of the effects of changing economic structures on science that we have been pursuing. But it is also the case that the reciprocal relation, from science to economy, is also at work. It follows that one cannot properly understand the *economic* aspects of the economics of science and the trajectory of change of the funding structures of science in isolation from consideration of developments in the sciences they are funding, as the latter is a crucial and irreducible moment in the seamless and indissoluble but messy feedback loops involved in the complex whole. The 'economics of science' and the 'economics of scientific knowledge' are thus merely two sides of the same coin.

Finally, there are at least two obvious respects in which shifting towards a co-production conception of science radically challenges the mainstream conception of economics. First, the very subject matter of an economics that is relevant to the study of the economics of *science* (as opposed to the various reified definitions of Economic Whigs and Mertonian Tories) demands that we employ an economics that is capable of exploring the *economy* as a socio-political – and, indeed, cultural – assemblage. In shorthand, we may say that this, at the very least, demands that a political economy perspective be assumed. Yet, such concerns have effectively been altogether banished from mainstream economics, to the point where many, if not most, economics undergraduates are exposed to no political economy at all – a state of affairs causing increasing dissatisfaction that has erupted in the disciplinary crisis mentioned above.

The second challenge is methodological and arguably more profound. Since co-production posits a social ontology of science in which the very nature of science develops alongside that of its broader socio-economic context, it becomes epistemologically impossible to employ a framework that must first define what science is before proceeding to examine its economics. Yet, this is precisely what the mainstream conception of economics does demand. Whereas the co-production analysis is thus concerned to *develop* our understanding of the nature of science(s) through analysis of its interactive development with is socioeconomic context – i.e. to stretch towards a 'definition' of science as its *conclusion* so that 'what is science?' is an open question for the economics of science – the axiomatic and 'deductivist' structure of mainstream economics requires the 'science' it is investigating be defined *ex ante*, at the outset of its research. This, in turn, leads to the reification of science and we have already considered above the numerous further problems that arise from this.

Taken together, therefore, these two challenges illustrate how an economics of science offers a singular opportunity and motivation for a broader project to develop an alternative economics that breaks with the mainstream discipline. For such a research programme addresses issues that will be at the very heart of economic concerns for future generations (cf. Sent 1999 for similar sentiments). Quite simply, *mainstream economics cannot illuminate the commercialisation of science and the knowledge economy more broadly*, and demand for just such understanding can only grow, especially in the context of crisis and discontent such as the present. Yet, as we shall see in Section III, it has been known for years that mainstream economics is 'broke'. Such is its dominance over economic enquiry, however, that any amount of demonstration can, and has, simply been dismissed as so much empty verbiage – the idle chatter of armchair methodological critics (who 'can't keep up with the maths'), while the 'big boys' get on with the 'real work' of economics itself.[16] Nor will such dominance be readily relinquished, even in the teeth of its evident intellectual bankruptcy.[17] As such, only when an alternative framework of compelling cogency has emerged, directly challenging mainstream economics in the 'real work' of economic analysis, will there be any chance of a substantive move away from this zombie social 'science' (Fine 2010). As Buckminster Fuller puts it (quoted in Dennis and Urry 2009: 9): 'You never change anything by fighting the existing reality. To change something, build a new model that makes the existing model obsolete.' The economics of science is perhaps uniquely placed to offer just such a prospect.

1.6 An alternative approach

We have seen, therefore, that the economics of science *cannot* be a mainstream economic project if it is to be an 'economics' and of 'science'. Several objections, however, are immediately apparent. First, regarding the methodological challenge of a non-axiomatic economics of science, it may rejoined that this is all very well, but how else *can* you proceed other than to start off with explicit assumptions about 'science', at least so long as the research programme is to be a social *science* and not merely anecdotal conjecture? Second, while the considerations regarding both 'economics' and 'science' point clearly in the direction of a political economy of science, what form should this take and why? Gesticulating to 'political economy' is just to open up a further set of questions, not to propose a pat solution. Similarly, has not the proposed alternative of co-production itself been too brief, so that the question remains, why is this position justified? In short, what is proposed instead?

In this book, we shall assume a relational and systemic approach to the economics of science. Accordingly, while we have seen that trying to model 'science' and its interaction with 'the economy' is precisely the *wrong* way to proceed, neither can these concepts simply be dispensed with. A *starting* point, therefore, is to begin by conceptualising the broader interaction of these two phenomena, where both are understood as emergent abstract realities in the process of ongoing (re)construction that each conditions the development of the other. However, both 'science' and 'economy' must immediately be expanded along at least two dimensions.[18] First, there is the distinction between the 'material' and 'ideational' aspects of both terms; for example, the laboratories and apparatus versus the theories and arguments

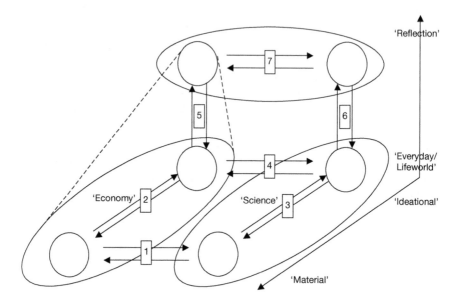

FIGURE 1.5 The economics of science (I)

of science, or the ships, factories, computers and machinery versus the contracts, bargains and regulations of the economy. Second, the ideational elements must themselves be separated into the formal or reflective 'scientific' ideas and the everyday or lifeworld ideas at work in these social 'spheres'.

As Figure 1.5 illustrates, we thus already have a systemic picture for an economics of science that is very much more complex than that of the NES. For instance, on this systemic approach, given that all these relations are part of the same system and so all affect each other either directly or indirectly, the economics of science must incorporate:

1   the contribution of science to new economic objects, most obviously as technologies and, conversely, the economic production of scientific apparatus, instruments and reagents;
2   the mutual conditioning of economic 'material' realities and economic ideas, visions, imaginaries and expectations;
3   the equivalent mutual conditioning of scientific technological, institutional and infrastructural realities and scientific theories and empirical data;
4   the mutual interaction of scientific knowledge and everyday understanding in economic activity, including the (political) economic impacts on the directions of scientific thought, argument and controversy;
5   the potentially performative and/or ideological effects of formal economic knowledge in and on the economy, and the reciprocal impacts of everyday economic understanding on framing economic enquiry;

6   the interaction between 'diurnal' and 'nocturnal', tacit and codified, practical 'wet lab' knowledge and theoretical reflection in scientific advance; and

7   the exchange of ideas between (natural) science and economics.

Moreover, the 'diagonal' relations between scientific materiality and economic ideas and economic materiality and scientific ideas could also plausibly be added.

Yet, this is by no means the end of this process of theoretical construction, for at least a further five steps also need to be taken, each adding commensurately to the number and complexity of relevant relations for an economics of science. We will introduce the last two of these below, but the first three may be mentioned here. First, proceeding further up the 'reflection' axis, we must add the scientific reflection on science itself, the discipline that is now often called 'science studies' or, more generally, 'science and technology studies' (STS). Second, a further 'plane' must be added in the form of 'nature' on which the social activity of both economy and science entirely depends (rendering 'nature' more accurately as 'socionature'), which economic activity relentlessly shapes, uses and exploits and which scientific activity (i.e. '*natural* science') takes as it object of inquiry. Indeed, the inclusion of 'nature', alongside 'economy' and 'science', as multiple and inseparable dimensions of an economics of science illustrates and reflects the diverse ways in which science and its funding are inseparable from (modernist) concepts of human progress: respectively the control of nature, scientific, rational mastery and (apparently endless) economic growth (e.g. Stirling 2009). This, in turn, reminds us of a key issue conditioning the whole of the diagram, namely the crucial interaction of conceptions of progress embedded in the 'economics of science' per se and the rest of the diagram. In other words, and third, given the systemic approach taken, it must also be acknowledged that, just as the scientific study of science sits *within* this system, so *too does the economics of science itself*. As we shall see, recognising this positioning of the epistemic project of an economics of science as *within* the reality it is also studying leads one to re-conceptualise the purpose and role of any such examination away from that implicit in the NES and other mainstream projects.

Having introduced this systemic conception of an economics of science, however, we must immediately qualify the diagram with a number of key disclaimers regarding how it should and should not be interpreted. First, it should be emphasised again that the diagram is (at best) only a rough starting place to illustrate the complexity of an economics of science and to direct possible lines of enquiry. In short, the diagram is to be used to go *beyond* it (and the inherent limitations of a static representation in two dimensions) not to be filled in and completed (whatever that may mean); to situate and connect different bodies of literature that are relevant to an economics of science but perhaps currently conducted in mutual isolation; and to *problematise* as well as open up and so assist in *that particular way* the totalising synthesis of ideas about the economics of science.

Second, the question arises of how the 'spheres' and 'relations' depicted are to be interpreted. As we shall argue at length in later chapters, the answer to this question must be in terms of the messy mediation of constitutively relational but emergent phenomena, shot through with incompleteness and uncertainty. Hence the diagram does not depict ahistorical and strictly distinguishable social 'spheres', nor even are the purported dualisms that set up the axes of 'material' versus 'ideational' or 'lifeworld' versus 'reflection' to be upheld without considerable qualification and deconstruction, especially in the present context when 'the gap between knowledge and action has been narrowed to such an extent that they cannot be separated, temporally or organizationally' (Nowotny *et al.* 2002: 183). The diagram thus raises as a *question*, rather than taking as given, the socio-historical *emergence* and ongoing (re)constitution of real abstract totalities that may credibly be called 'science' and/or the 'economy'. Similarly, the relations should not be understood as formal but as substantive, hence manifest in particular instances and so demanding detailed empirical attention, so that the diagram should more properly represent 'sciences', 'economies', and '(socio-)natures'. Indeed, this leads to what may be called a 'double challenge' for the economics of science, as needing simultaneously to offer such abstracted insights regarding the mutual mediation of 'science' and 'economy' as are forthcoming *and* to attend adequately to the concrete particularities of actual instances, which in turn raises key questions for further abstract and theoretical reflection. For instance, it is through detailed empirical analysis of the new biotechnologies or genomics that key issues regarding the hopes, promises and fears of these new technological capabilities, and investment in them, arises.

Finally, the diagram should be understood to direct inquiry to *dynamic* processes (i.e. adding a time dimension), in which the focus is on the parallel and interactive co-production of (cultural, political and economic) socio-technical and techno-scientific trajectories of change. Moreover, taken together with the substantive focus just discussed, this also means that 'history matters', so that the economics of science should examine actual historical processes and the multiple conditionings and contingencies these inevitably involve, leading to contingent path dependencies and systemic 'lock-in' and 'lock-out'.

Against the NES and other such projects, therefore, the overwhelming conclusion must be that there is *no single economics of science*, nor a single 'perfect' policy for the funding of science that can be expressed in ahistorical and purely abstract terms; a claim that once stated explicitly seems almost transparently absurd (cf. Minogue 1993). Nor, therefore, is formulation of any such policy the task for an economics of science, but rather a detailed and non-reductionistic examination of the relational and structural realities of particular sciences and national systems of scientific research in order to inform broad political debate about the directions of techno-scientific change. Indeed, a crucial aspect of this work is thus to show what (and how little) *is* known or even knowable, thereby problematising and *re*-politicising issues that prematurely universalistic frameworks such as the NES would serve only to obscure and close down (cf. Stirling 2005).

The diagram thus helps us conceptualise the purpose and task of an economics of science quite differently. The importance or usefulness of this initial systemic diagrammatic representation, however, is not just a matter of its potential assistance in situating and synthesising diverse work. It also serves to bring together seemingly diverse *issues*, thereby also heightening both the potential insights available and, at the very least, furnishing an even greater sense of urgency regarding the reality under investigation, taking us beyond purely epistemic concerns to the omnipresent and crucial question of all social inquiry: so what? It is clear that the enormous changes discussed above in the funding of science have themselves generated sufficient anxiety to motivate the flurry of interest in the 'economics of science' in recent years. But through this systemic perspective, by incorporating broader issues regarding science and knowledge, the economy (which, recall, is notably and paradoxically absent in most frameworks) and even 'nature' introduces a plethora of further issues, all of which are also of direct importance to an economics of science *and vice versa*. Moreover, there is a growing sense (and commensurately burgeoning literature) regarding each of these social phenomena themselves being currently in a state of *crisis* and in ways that interact in important ways with the commercialisation of science.

For instance, regarding science itself, numerous commentators have discussed the increasingly tight overlap and interaction between 'science' and 'society', as the very *success* of science – in developing new forms of technological intervention, the ubiquity of science and technology in everyday lifeworlds and an increasingly scientifically literate public – generates growing scepticism and contestation regarding particular techno-scientific developments and their associated proliferation of new risks and uncertainties (Beck 1992, 1999, Callon *et al.* 2009, Epstein 1996, Jasanoff 2005, Latour 2004, Nowotny *et al.* 2002). As such, the democratisation of scientific processes has become a crucial concern, both for science itself and for society more broadly (Jasanoff 2007, 2010, Leach *et al.* 2005, Stirling 2009, Wynne and Felt 2007), while policymakers and scientists, in particular, commonly (mis-) interpret this in terms of fears regarding a crisis in the diminishing authority of science and scientific expertise (Bijker *et al.* 2009, Collins and Evans 2002). Indeed, the challenges to the epistemic authority of a separate social institution designated '(the) science' is part of the broader challenges to the modernist worldview of diverse 'post-modernisms', multiculturalisms and cosmopolitanisms that constitutes 'reflexive modernization' (Beck 2006, Beck *et al.* 1994). Moreover, this process of 'society' increasingly 'speaking back' to 'science' (Nowotny *et al.* 2002) is compounded by the changes in the economics of science. For instance, the commercialisation of science – and thus the potential for private gain from techno-scientific innovation – frames and heightens public concern regarding the accountability and desirability of particular techno-scientific innovations (e.g. genetically modified (GM) crops, biobanks and biometrics, 'designer babies', stem cell treatments, cyber security, etc.), even as policy may attempt to neutralise political debate on these issues by attempting to 'scientize' it (Levidow *et al.* 2007).

Similarly, it is unarguable that since 2008 (and from much earlier on some accounts, see Section IV in Volume 2) the global political economy has been racked by financial and economic crisis, particularly in the 'core' of the rich global North. This crisis is compounded by the challenges of the rise of new economic powers, especially China, and the novel challenges arising from processes of globalisation. Again, the economics of science is surely central to these concerns. For instance, globalisation of economic activity increasingly includes innovation and R&D (Archibugi and Iammarino 2002, Dicken 2007, Ernst 2008) while ongoing attempts to forge a new 'techno-economic paradigm' of a 'knowledge economy', with the progressive collapse of both welfare state industrial capitalism since the 1970s and, perhaps now, the neoliberal, financialised order that has replaced it (Birch and Mykhnenko 2010), have focused overwhelmingly on uniquely science-intensive industries, such as biotechnology, nanotechnology and ICTs.

Finally, it is also widely accepted that we currently face numerous ecological crises of unprecedented severity and global scope, including climate change, biodiversity loss, soil erosion, resource over-exploitation and pollution (including water and deforestation) and novel pandemics. These challenges are themselves crucial aspects of the current economics of science, increasingly setting research agendas and techno-scientific trajectories of innovation, but they are also related to the economics of science in terms of the construction of these forces of destruction in the first place and, in the case of the new technologies in which so much hope and money is invested, in the generation of novel ontological challenges that unsettle the very category of 'nature' and the self-conceptualisation of human beings' place 'in the world'.

In short, therefore, the systemic diagram allows us to situate the programme of an economics of science in the much wider set of urgent challenges of this 'triple crisis', all three aspects of which – knowledge, political economy and ecology – converge in the central political discourse of the 'knowledge-based bio-economy' (KBBE) (see Chapters 2 and 3).

1.7 A philosophical detour?

So much, then, for an overview of this alternative approach. But questions of justification of this alternative perspective remain central; and, indeed, particularly intransigent in the context of an economics of science, in which questions of 'science' and its epistemic warrant arise continually, at multiple, recursive levels of reflexivity. It is, thus, at this point that we must turn to questions of philosophy on the basis that they, i.e. *philosophical* concerns, are always present and hence can either receive due attention, and so strive towards some sort of provisional theory-practice consistency, or go neglected, and then most likely linger as festering weaknesses. Thus, while the theoretical objections to a mainstream economics of science and further argument at that level regarding any proposed alternative are essential and indispensable, the radical inadequacy of existing 'economics of science' projects draws particular attention to crucial philosophical questions that must themselves be addressed, if only to identify, and so avoid, underlying philosophical problems with the whole paradigm of mainstream economics (and other relevant disciplines) as we get on with the more important task of *constructing* an alternative economics (of science).

As such, we must embark on something of a philosophical 'detour' in our quest for an economics of science. This may, however, seem an unpromising route to take: after all, is not the philosophy (of science) itself plagued by its own insoluble problems (e.g. Barnes 1974)? Will not straying into philosophy simply take us permanently away from the economics of science, into an entirely different 'wilderness' (e.g. Hands 1994, Tyfield 2008b, 2008c)? Such caution is certainly well founded, and vigilance is needed so as not to get lost in arcane and abstruse philosophical argument, often pursued apparently for its own sake. Yet, the argument of this book is that, these risks notwithstanding, one cannot have a legitimate social science without explicitly situating it within the philosophical reflection and reasoning upon which its justification depends. Equally, however, and against the grain of much of the philosophical literature, one cannot have productive and philosophy capable of rational judgement without a separate body of thought, such as social science, from which the questions worth addressing arise. Philosophy and (in our case) social science thus should be viewed as proceeding as relatively autonomous but mutually dependent parallel discourses, with at least part of their respective epistemic warrant always lying elsewhere.

This transformed picture of the relation of philosophical and social scientific work then, in turn, sponsors a particular set of answers to substantive philosophical questions (of ontology, epistemology and methodology) that provide both grounds of justification for particular social scientific theoretical perspectives and thus a reflexive and dynamic developmental consistency between philosophical and social scientific positions.[19] In particular, this book argues that the philosophy of 'critical realism', based on the seminal work of Roy Bhaskar, Tony Lawson, Margaret Archer and Andrew Sayer, is singularly useful in this regard because of its unique concern with the examination of ontological presuppositions as *the* pre-eminent form of specifically *philosophical* reasoning. Critical realism (CR) thus deploys a methodology of immanent critique to explore the presuppositions of stated positions to uncover what must be the case for that position to be intelligible. As such, it primarily works 'backwards' or 'downwards' by processes of retro- or ab-duction, including by way of transcendental argument (TA), rather than deductively from axioms or inductively from empirical observation of patterns; without, of course, dispensing with these essential modes of reasoning. In this way, it also illuminates presuppositions that may themselves be 'real', in terms of having causal effect in the world, as the necessary conditions of intelligibility of other causally effective and acted upon beliefs. This therefore adds the fourth element to our diagram, namely the importance of the distinction and relations between explicit beliefs and tacitly presupposed commitments, *both* of which may have real causal effect.

This approach thus offers a non-foundational, dynamic and fallible, but also distinctively ontological, grounding to social scientific theory, exemplified in concepts of contingent and conditional natural necessity, as well as a particular understanding of the place, purpose and method of philosophy and social science alike as simultaneously and inseparably *critical* (in the Kantian sense, inter alia) and *realist* endeavours. In this way, then, an entirely different approach to mainstream economics is afforded and elaborated, answering the two sets of objections above regarding non-axiomatic social science and justification of alternative perspectives. Indeed, in the former case, non-axiomatic social science is not merely shown to be possible but also justified in a way that axiomatic social science is not, at least without grounding in the former.

It is for this reason primarily, therefore, that this book sets out a *critical realist* overview of the economics of science. But a further five sets of advantages from assuming a critical realist stance may be identified regarding this particular subject matter. First, the core ontological argument of critical realism affords the elaboration of a unique ontology that is both attentive to (rather than dismissive of) the concept and nature of 'reality' but also attuned to the constructivism constitutive of human or social realities (including diverse socio-natures) (e.g. Benton 2001). The ontology of critical realism thus conceives of 'reality' as simultaneously real and conditional, emergent and mediated, structural or systemic and relational.

Critical realism's concern with ontology also thereby sponsors an entirely different epistemology from that of most mainstream philosophy of science (and sociological criticisms of it) in that the social construction of knowledge (or 'epistemic relativity') is understood as *constitutive* of science and an essential precondition of (the possibility of) judgements of truth (or 'judgemental rationality'), rather than as an intractable circle to be squared. As such, critical realism furnishes an ontological understanding, i.e. regarding the nature of 'reality' per se rather than a list of its constituents, that does not posit nor depend upon a 'hard core' of (meta-)physical knowledge; an intractably problematic presumption for sociological analysis and especially of the changing role of scientific knowledge in social change (e.g. Nowotny *et al.* 2002: 192 ff.). Similarly, critical realist epistemology legitimates both the socio-historical constructivism (and hence anti-transcendent scepticism) of scientific knowledge *and* the *possibility of* provisionally conclusive and representationally adequate truth; a key issue, for instance, for critical analysis of the inadequacy of crudely 'positivist' framings of knowledge about nature and their real effects in terms of ecological destruction (Benton 2001, 2003).

Second, at the level of social scientific theory choice, assuming a critical realist perspective regarding the economics of science in particular sponsors a relational Marxist approach matched with analysis of the co-production of 'science' and 'society'. As such, the setting out of this theoretical argument (in Sections IV and V in Volume 2) takes us step-by-step from the 'economics of science', to a *'political* economy of science', to a 'political economy of *research and innovation'* (R&I), to a *'cultural* political economy of research and innovation' (CPERI). Moreover, critical realism thereby provides the philosophical grounds on which to build productive synthesis of what are otherwise foundationally incompatible disciplines regarding the 'economy' and 'science', as is needed for an 'economics of science'. For whereas political economy is concerned with real social structures that are credible only in the context of a philosophical realism, the discipline of STS has been founded (and most famously and fruitfully developed) on an essential repudiation of realism, including of real social structures. This reciprocal challenge, of 'economics' to 'science' and vice versa is then evidenced by the relatively poor attention mutually paid between these disciplines to date, despite their prima facie relevance in both cases to the same project of an 'economics of science'.

Grounded in a critical realist philosophy, which is both ontologically monist and analytically dualist, however, productive engagement becomes possible, opening up the reciprocal examination of the conditioning of techno-scientific trajectories by political economic structures, on the one hand, and the mediation of the construction of economic value and the regularisation of capital accumulation by techno-scientific innovations, imaginaries, institutions and materialities, on the other. A relational Marxist perspective thereby affords an analysis of scientific change that includes the reality of the social relation of capital as a key but neglected, and necessary but insufficient, element of non-reductionistic explanation of these crucial, causally over-determined processes.

Thus, while a distinctive contemporary move within those social sciences related to issues of an economics of science is towards a relational methodology and ontology (e.g. in STS (Law 1991), rural sociology and agrarian studies (Goodman 2001), environmental sociology and/or geography (Castree 2002), economic sociology including studies of globalisation (Gereffi *et al.* 2005) or economic geography (Thrift 2005)), a critical realist approach affords a theoretical perspective that both resonates with these developments *and* rectifies the persistent absence in much of this work of enduring structural issues, always already there at any given point from which to start empirical research, not least through its ability to incorporate real abstractions as non-foundational presuppositions of meso-level relational analysis.[20]

Substantively, analysis in these terms also brings out a singularly important conclusion, namely that current attempts to construct a 'knowledge-based bio-economy' must be understood, at least in part, as a neoliberal project, incorporating diverse and contradictory individual and institutional agency (including of the *state*), to expand the colonisation of social life by capital into new social realities, namely (scientific) 'knowledge' and 'nature'. This then places examination of the likely effects of this attempted enclosure, its abstract tendencies, immanent contradictions and concrete manifestations, as a key question for an economics of science. This is also, then, the final element that we want to add to our diagram, namely the asymmetrical expansion and growing overlap of 'economy' and 'science', allowing us to 'complete' the diagram (provisionally, of course) as in Figure 1.6.

Finally, and most broadly, a critical realist perspective sponsors a deflationist philosophical perspective (Sismondo 2007), in which the immanent critical method highlights the inherent limitations of knowledge – and more generally of associated forms of social life dominated by 'cognitivist' social understandings of knowledge – but without entirely forsaking or repudiating the gains and potential of scientific knowledge(s). In particular, while the relational and constructivist aspects of critical realist philosophy of science emphasise the inherent inadequacy of assuming a realist stance, the transcendental and critical aspects stress the impossibility of merely repudiating such realism altogether. A critical realist perspective thus reveals and

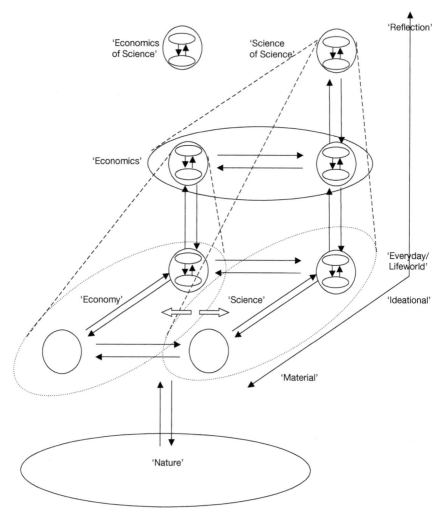

FIGURE 1.6 The economics of science (II)[21]

focuses on the intrinsic epistemic *predicament* of contemporary humanity, both individually and collectively (and hence of social life per se), as the *inadequate but inescapable* presumption of the reality of *some* beliefs (hence some *realisms*) at any given time. This is, therefore, to counsel an unshakeable burden of humility regarding both the limits *and the possibility, reality and normative obligations* of knowledge. Against the 'untenable . . . persistence of an apparently single unitary mainstream normative understanding of progress' (Stirling 2009: 8), therefore, this in turn conditions a reappraisal not only of the purposes of the economics of science itself, as a scientific project, but also and more profoundly of the role, purpose and politics of knowledge in society more broadly.

Resonating with this worldview, then, the particular goal of an economics of science/scientific knowledge is *not* to furnish a relatively certain and fixed body of theoretical knowledge that affords accurate prediction and policy manipulation of the knowledge economy in order to maximise the allocative efficiency of science. Rather, it aims to engage in active collaboration with diverse social actors as a transformative praxis, in order to contribute to and to *(re-)direct* the ongoing trajectories of socio-economic and techno-scientific change, including by making explicit the socio-political processes causally affecting them and the actual limitations of present knowledge of these processes for the purposes of public political contestation and participation.

1.8 Dialogue with critical realism

This is not, however, to claim that the philosophy of critical realism is the solution to all the various problems with an economics of science, let alone the 'triple crisis' with which it must grapple. For starters, it is clear that social science is a relatively autonomous discourse and practice that must take place on its own level (as is social action itself). Philosophical argument, therefore, can neither legislate for nor replace it. It follows that there is really no such thing as a 'critical realist economics', though this is emphatically not to say that critical realism has no impact on what forms of economic analysis are justified (see Section III). Indeed, a common criticism of critical realism as a school of thought is that it dwells only and ever in philosophical argument and never shows the difference it makes at the level of concrete theory. Responding to and rebutting this criticism is a major goal of this work, even while it is conceded that there has been insufficient effort within critical realism to take this step, if not (by some influential critical realists) positive effort to rule it out.

Second, as for all bodies of thought, critical realism is a relatively loose, if well-formed, assemblage of ideas that are themselves in development in response to the wider set of philosophical and theoretical argument to which they are responding. It is thus, inevitably, imperfect if not flawed, inchoate and unhelpful in some regards; in short, a living philosophical research programme, not a body of doctrine (Bhaskar 1986, Jessop 2002b). As such, there is plenty of room for improvement in critical realism as it evolves *alongside* the developing social scientific theory that is the material for its reflection. In particular, while critical realism fits quite easily with Marxist and other heterodox political economy, hence a large and growing literature in this vein, the book will focus on the mutually productive interaction and tension with STS; an engagement that is particularly productive precisely because of their differences, as well as their overlaps.[22]

Third, therefore, given that philosophy and social science are conceived as parallel discourses, the interaction of these two must be considered *in both directions*. If demonstrating the effect on social science of assuming a critical realist philosophical stance is rare, this other moment of the 'dialectical' relation of exploring how critical realism *itself* is transformed by interaction with social science (perhaps inspired by other philosophical commitments) is more or less neglected altogether. Hence, this is the second major gap in the literature of critical realism and, again, one this book aims to address to some extent. The book is thus an exploration of the economics of science *in dialogue with* critical realism, i.e. a parallel investigation of social science and philosophy, just as is advocated.

In particular, the following criticisms or weaknesses of critical realism, or ways in which more concerted engagement with pragmatist or constructivist work would benefit critical realism, emerge through the book. First, there is the productive cross-fertilisation of critical realist argument and critical realism-inspired theory from greater attention to empirical analysis and concrete case studies – a 'critical realism in concrete' (per Jessop 2002b) – which is all too often lacking in critical realist literature. Such work would employ a variety of methods (e.g. Downward 2004) to triangulate and illuminate the diverse processes conditioning phenomena of interest. Indeed, while much of the discussion in this book focuses on the modes of transcendental argument (TA) and the immanent analysis of Marxian critique of political economy, this is because these represent the singular contributions of a critical realist perspective and those neglected by other approaches. Indeed, it would be a complete misreading of critical realism more generally to interpret it as the argument that these are the only relevant forms of analysis, or that capital is the only relevant factor in any social scientific explanation.

Nevertheless, engagement with the empirical and constructivist work of STS does offer some important lessons for critical realism more generally. In particular, Bhaskar's seminal argument for a depth or transcendental realism (see Section III) is too often interpreted by critical realists as *conclusively* establishing an account of the nature of reality (i.e. an ontology) and associated definition of science that is effectively a secure starting place for all further enquiry. Conversely, concerted engagement with constructivist social studies of actual scientific practice serves to loosen, relativise, mess up and situate such foundational commitments, including the nature and warrant of the transcendental argument itself, and in ways that are as a result both epistemologically and normatively more reasonable and appealing. For instance, critical realist ontological argument tends to underplay the sheer contingency of the emergence of the constituents of reality, as well as the human construction of novel (especially scientific) phenomena, thereby neglecting a hugely important qualification to understanding of the *ontological* significance of contemporary science, even as it claims to offer the best philosophical account of such knowledge.

Similarly, while critical realism itself has been developed, particularly by Bhaskar, as the unfolding of a philosophical project through a recursive and dialectical process of transcendental analysis to ever 'deeper' levels of ontological analysis, this 'vertical' and a priori investigation, for all its insights, has also come at the cost of acknowledgement of the limitations and fallibility of such conclusions and their practical or political implications. For instance, while a compelling argument is presented for the derivation of facts from values and vice versa, by way of an 'explanatory critique' (Bhaskar 1986, 1998), thereby apparently affording scientifically grounded critique, the very conclusiveness of such argument is too quickly understood to have dispensed with the need to subject any such normative conclusions themselves to further argument; for instance, regarding the resulting *axiological* imperative thereby to act to demolish the actuality thereby judged to be normatively negative and/or *comparative* normative appeal of that actuality vis-à-vis possible alternative realities (e.g. Sayer 2000). Critical realism also thereby deploys a concept of 'critique' and critical social science that is too easily taken as an exhaustive substitute for messy, strategic and concerted political action.

In short, the fundamental challenge for critical realism is not to yield to strong and evidenced tendencies, as a real*ism*, to interpret its task (if not singular mission and destiny) in purely or primarily cognitivist terms, i.e. to re-establish on surer ground the cognitive authority and ability of (albeit a different, critical) science (and social science and philosophy), which has been so beset by the multiple challenges of post-modernism (Nowotny *et al.* 2002: 186–187) and the intrinsic inadequacies of a crude 'positivistic' realism. Instead, the task for critical realism is to play a part in *resituating* such authority and limitations of knowledge and to move, in reality and in practice, beyond these outdated conceptions towards diverse, sustainable and equitable learning societies: in new models of scientific research (i.e. as multidisciplinary, international, democratically-engaged collaborations) and practices of knowledge production that challenge the capitalist subsumption of 'knowledge' and 'nature'; and in forms of social and individual self-transformation so as to be able to respond positively to and develop these new realities.

Yet, in all these respects, it may be argued that this is not to repudiate a critical realist perspective but rather to remain faithful to its essential 'realistic spirit' (cf. Diamond 1991) in which reality always hugely exceeds knowledge, the consequent inseparability of its 'critical' and 'realist' aspects and the inherent, but endlessly discomforting, imperfectability of knowledge that this entails. As such, this remains a 'critical realist overview' because, despite the weaknesses just outlined, critical realism remains a singularly important (but alone 'insufficient') conceptual resource, and one that has as much, if not more, to offer constructivist STS in return.

To conclude, then, this book has three aims:

1   First, to illustrate the alternative conception of an economics of science introduced schematically in this chapter in the context of critical and empirical explanation of the contemporary commercialisation of scientific knowledge production, together with discussion of a broader substantive agenda of crucial contemporary importance to which an economics of science should be making important contributions.

2   Second, to set out and justify this alternative research agenda, insofar as is possible given the 'double challenge' of both (i) giving due attention to empirical differences and concrete detail and (ii) setting out such abstracted insights regarding the contemporary 'cultural political economy of research and innovation' as are available. In doing so, therefore, we provide some illustrations of the former, attending to the increasingly stark limitations of knowledge about

knowledge and techno-scientific uncertainty upon which arguments of 'Mode 2' science and society (Gibbons *et al.* 1994, Nowotny *et al.* 2002), 'cosmopolitics' (Callon *et al.* 2009, Latour 2004, Stengers 2010) and the sociology of ignorance (Gross 2010) have focused; while *also*, regarding the latter, showing what critical social science can still uniquely provide.

3   And third, to illustrate the mutually productive engagement of critical realism and constructivist studies of science and at multiple levels: ontologically, epistemologically, methodologically and politically.

Chapter plan

The book, divided into two volumes, proceeds as follows:

In **Section II**, we begin with the examination of the substantive 'economics of science' and its changes, employing the perspective described in brief above. This serves two purposes: to furnish substantial insights on the subject matter and thereby motivate the extended theoretical and philosophical discussion of the rest of the book in terms of what is at stake; and second, to illustrate the explanatory power of that theoretical perspective in terms of its capacity for such insights. Given the impossibility of setting out a comprehensive 'economics of science' that covers all relevant actual cases, the chapters provide a selective overview of a number of crucial trends, namely the 'knowledge-based bio-economy' (Chapters 2 and 3), global IPRs (Chapter 4) and the commercialisation of Chinese science (Chapter 5). We thus also trace a rough route from the most abstract characterisation of the contemporary knowledge economy to more concrete discussion of a particular example of the commercialisation of science; from the global to the local and from the EU, to the US, to China; and from post hoc explanation of globally dominant developments to exploration of potentially key future trends.

In **Section III**, we then begin our journey in rebuilding and justifying the theoretical edifice deployed in Section II. True to the argument outlined above, our exposition begins with an articulation of and critical engagement with the critical realist philosophy of science, and its critique of mainstream economics in particular, given our target of an 'economics' of science. In so doing, we also qualify and go beyond established critical realist argument; in particular, through engagement with constructivist STS, leading to a development of critical realism that may be called 'transcendental constructivism'. Similarly, in explaining the core methodological insight of critical realism regarding the realist transcendental argument, we also contextualise and situate this form of reasoning and the ontological and epistemological conclusions to which it leads.

Volume 1 thus offers the 'top' and 'tail' of the argument, regarding the substantive and philosophical aspects of the argument. In Volume 2, we then move on to consider the theoretical building block of an alternative critical and explanatory economics of science that joins the two. This takes the form of the reciprocal disciplinary challenges of 'economics' and 'science' and the capacity of a critical realist perspective to facilitate their productive (but never complete) synthesis. Hence, in **Section IV** we consider the contribution and limitations of STS vis-à-vis an economics of science, while in **Section V** we turn to political economy and the evolutionary economics of innovation (EEI).

Finally, we conclude with reflections regarding 'what next?': for the research programme of an 'economics of science', or rather a cultural political economy of research and innovation (CPERI); for the political engagement of such research; and for the real challenges of the 'triple crisis' themselves.

Further reading

David, P. and P. Dasgupta (1994) 'Toward a New Economics of Science', *Research Policy* 23: 487–521.

Foray, D. (2004) *The Economics of Knowledge*, Cambridge, MA: MIT Press.

Mirowski. P. and E.-M. Sent (eds) (2002b) *Science Bought and Sold*, Chicago: University of Chicago Press.

Mirowski, P. and E.-M. Sent (2008) 'The Commercialization of Science and the Response of STS', in E. Hackett, J. Wacjman, O. Amsterdamska and M. Lynch (eds), *New Handbook of STS*, Cambridge, MA: MIT Press.

Radder, H. (ed.) (2010) *The Commodification of Academic Research: Science and the Modern University*, Pittsburgh, PA: University of Pittsburgh Press.

SECTION II

The commercialisation of science and the construction of the knowledge-based bio-economy

2

THE KNOWLEDGE-BASED BIO-ECONOMY

Knowledge and nature as sources of value?

2.1 What is the KB(B)E?

A major, if not the overwhelming, motivation for the surge in interest in an 'economics of science' since the late 1990s has been the seeming emergence of a new economy in which knowledge, including science, is assuming pre-eminent importance. A central question for an economics of science is thus: 'what is the knowledge-based economy (KBE)?'

First, the KBE refers to what appears to be a number of secular trends in the organisation of economic activity towards (Powell and Snellman 2004): increasing economic resources devoted to knowledge production; increasingly science-intensive innovation; a qualitative shift in the distribution of knowledge production capacity towards user-led innovation (von Hippel 2005); and the growing importance of 'integrative knowledge', which is needed to be able to manage the quantum growth in knowledge production (Foray 2004).

More recently, however, and of equal if not greater importance for an economics of science, the KBE discourse has developed under the expanded title of the 'knowledge-based bio-economy' (KBBE). The KBBE was at the heart of the European Union's (EU) Lisbon Agenda to become the most competitive economic bloc in the world by 2010 (Euractiv 2004a, 2004b, European Commission 2000, 2007, Rodrigues 2003). Subsequently, as a major EU policy discourse, it has taken centre stage in shaping scientific research agendas, with 'KBBE' themes and agendas featuring prominently in the latest round of its major funding programme, the Framework Programme (Framework Programme 7 2010). Building on existing policy understanding of the importance of 'knowledge' in the new economy, and thus to European economic competitiveness, these funding schemes are based on EU analysis of the importance of the European 'bio-economy' in at least two respects: first, as an existing and growing economic sector, already worth an estimated €1.5 trillion (Directorate-General for Research 2005), and with singular potential for further innovation and associated competitive advantages; and second, as a direction of economic development that promises the squaring of economic growth with the exigency of the current environmental crisis/crises (European Commission 2002, 2010, Directorate-General for Enterprise 2009). Moreover, the reach of equivalent policy visions or 'imaginaries' extends way beyond European governmental and quasi-governmental institutions. This would not only include policies from elsewhere (importantly including the US) that deploy similar, if not identical, terminology (e.g. OECD 1995, 2005). It would also include the even more diffuse and inchoate imaginaries of the future structure of the economy that enervate and underlie strategic and tactical decisions for businesses and researchers across the world.

Both KBE and KBBE discourses thus are characterised by two major claims. These two claims are: (1) that knowledge and (for KBBE) nature are emerging as the primary sources of value in economic activity; and (2) that this shift in value creation prefigures a wholesale transformation of the economy towards the goals of enhanced resource efficiency, reduced ecological impact and environmental sustainability. This latter claim is then usually justified in terms of the reduced resource intensity of an economy based on knowledge, with the possibilities of 'dematerialisation' or 'decoupling' of economic growth and resource use, and a smarter and more ecologically-friendly impact of a bio-economy in which the autopoietic characteristics of 'natural' (or rather biological) systems affords limitless, recyclable and hence sustainable resource use (cf. Cooper 2008); what Birch *et al.* (2010) have called 'sustainable capital'. These two claims are largely invoked by proponents of the KBBE agenda, but a number of critical responses also accept the incorporation of knowledge and nature as sources of economic value. Against the utopian vision of proponents, however, such critics see these as disastrous developments, presaging the increased exploitation of these resources (i.e. knowledge and nature) and a supposedly endless and total colonisation of 'life itself' by capitalist imperatives (e.g. Brennan 2000, Cooper 2008, Rose 2007a, 2007b, and see also Sunder Rajan 2003, 2006).

This chapter engages with this core development for an economics of science by comparing the KBBE policy imaginary and its current and likely future (tendential) reality. In particular, we will broach two core questions. First (in this chapter) we will ask:

1. How is value actually produced? And, in particular, can knowledge and/or nature be a source of value?

Against this foundational claim of KB(B)E discourse, I argue that neither know-ledge nor nature can be sources of value; or rather, to be precise, that knowledge *labour* can be under some conditions but it is problematic to establish these conditions, they are intrinsically unstable and, where successful, always dependent on extra-economic (including anti- not just extra-market (cf. Albo 2007)) conditions (including, but not limited to, state regulation). Moreover, since the resolution of these tensions regarding knowledge labour as a source of value is impossible in the abstract, it can only ever be achieved temporarily and in concrete. It follows that, against KB(B)E discourse, there is no single path of economic development of and towards a KB(B)E, derivable in the abstract. Instead, this process depends on socio-historically and geographically particular political/cultural/technological conditions. Hence, explanation of the concrete, historical trajectory of economic change in particular places is an empirical question and (as discussed in the Introduction), there is no single and ahistorical economics of science.

With the KBBE discourse thus deconstructed, however, the temptation may be to leap to the conclusion that the KBBE is just empty buzzword (Godin 2006). Before we can accept this conclusion, however, we must explore the real place of knowledge and nature in KBBE. Hence, in Chapter 3 we turn to the second question:

2. What is the actual, existing KBBE?

Unable to discuss these issues at any length in the abstract alone, we will illustrate an answer to these questions with the key 'bio-economy' example of the globalised agri-food system. In doing so, I will argue that, first, the KBBE discourse is by no means merely an empty gloss but a crucial aspect of the ongoing attempts to construct it. In order to understand the particular power vested in this discourse, however, we must consider its structural support and in particular its resonance with the neoliberal financialisation of the global economy. This has underpinned an ongoing political project to 'neoliberalise' (agricultural) nature(s) in parallel with and with the assistance of a commensurate commercialisation of scientific knowledge production. By exploring some of the resulting concrete realities and their associated problems, we will then proceed to argue that, far from resolving problems of ecological sustainability, KBBE policy discourse is exacerbating both a global agrarian (and latterly food) crisis and the environmental crises about which it makes such grand claims.[1] Hence, we will also dispute the second core claim of KBBE discourse, regarding its (ecological, but also socioeconomic) sustainability. Finally, in the process, we will also exemplify two of the key points regarding the economics of science more generally, namely (1) to illustrate a critical and explanatory economics of science that must also grapple with the 'double challenge' of addressing both abstract generalities and concrete detail of particular 'economies' of 'science'; and (2) to show how the limited question of economics of science regarding (the supposedly optimal arrangement for the) commercialisation of science must be set in the context of wider and concrete changes.

2.2 Value as a term of art

Our analysis of the economy starts (following Marx's lead in the opening sections of *Capital*, volume 1) from the immanent perspective of acknowledging the particular form assumed by prosperity in current socio-economic circumstances, namely as an abundance of commodities. A commodity is a good (product or service) produced specifically to be sold. This entails immediately that the commodity has both a use-value and an exchange-value. The ubiquity of commodities, however, along with the existence of *systemic* exchange values (hence the 'market' in the abstract sense often invoked) shows that commodities are not just one among many of the forms inhabiting socio-economic life (e.g. along with gift or barter) but are the dominant such form. The starting point of our analysis, therefore, is the system of generalised commodity exchange (GCE). The analysis then proceeds by asking (Fine and Saad-Filho 2004: 17)*:* 'what is presupposed by the existence of systemic markets that give rise to the fixed ratios (or "exchange values") characteristic of commodities?' The answer to this question is 'value'.[2]

We may note immediately, therefore, that it is absolutely clear that 'value' (and its cognate of 'surplus value') is both a concept that is presupposed by actual, current economic praxis – and so one without which any analysis of the economy is effectively impossible – but also a *term of art* entirely divorced from the usual rich and polysemous connotations of the word (including normative approval or relating to a really-produced thing servicing genuine human needs). 'Value' is 'value *for and from the perspective of* capital', which, if anything, effects a trans-valuation of values that imposes a one-eyed, primary norm of profit accumulation above the concerns and needs of the social and natural systems on which economic life, including capitalist accumulation, parasitically depends; hence prioritising exchange-value over use-value, even as the former depends upon the latter. Using the terminology of 'value' in rigorous economic analysis, therefore, demands a shift in mindset to an altogether different and unfamiliar, indeed counter-intuitive, 'meaning universe'. When set against the ease and familiarity of talking about economic 'value' in everyday parlance, this is therefore to counsel the utmost caution whenever one reads or uses the term 'value' in an economic discussion.

Moreover, this 'value' (and a surplus of it, hence generating a profit that motivates capitalist production in the first place) is produced only by the consumption of a particular commodity in a production process (human) labour power. For consumption of the means of production, as inert participants in the production process, cannot impart value that is any more than that which went into *its* production in turn. Were this not the case, then exchange at value – which, recall, is presumed *ex hypothesi* in a capitalist market economy – would be impossible because commodities could increase their value with the simple passage of time. Conversely, labour power is the only commodity, albeit a fictitious one (i.e. something that is treated as a 'commodity' but is not in fact produced in order to be sold), that *can* separate its (daily and generational) (re)production from its consumption while also integrating the former in the circulation of commodities (so that it can have an exchange-value in the first place).[3]

Economic 'value', therefore, is produced by labour and labour alone. But one further clarification is needed before we proceed to use this terminology to explore the KBBE, namely the distinction between labour that is and is not productive *for capital*. This follows directly from the definition of the singular position of labour power as source of value, for not all labour (i.e. concrete labour expended in day-to-day socio-economic life) is productive of 'value', hence of *surplus* value, and so is productive *for and from the perspective of capital* (PLK). Rather, productive labour for capital (PLK) is defined by three criteria (Fine 1986, Mohun 2003, Savran and Tonak 2002), each of which is a necessary but insufficient condition, namely labour power expended in conditions of:

1   waged labour;
2   employed by capital (i.e. with a view to making a profit); and
3   in the sphere of production (of commodities, whether products or services).

We will return to this analysis of PLK in Chapter 3. But first note that this Marxian argument can, therefore, be described as a 'labour theory of value' (LTV); for example, that labour (power) is the sole source of value (for capital).[4] But, following Elson (1979), we may just as correctly describe it as a 'value theory of labour' (VTL), where we acknowledge that 'value' here is a term of art and hence the theory is an exploration of the implications of the social organisation of economic activity that treats labour *as if* it could be valorised, i.e. as if it were a commodity. Certainly, this latter approach has the benefit of undercutting the legion misunderstandings of Marx's account of value. Our first target, however, is the twin claims that nature and/or knowledge are (increasingly important) sources of value. If PLK alone is a source of economic value, we can see immediately that the central argument of KBBE discourse, whether boosterism or jeremiad, is false. But let us consider each of these claims in turn in more detail.

2.3 Nature as source of value?

First, why can't nature be productive of 'value'? For instance, Brennan (2000) argues that Marx's distinction between means of production and labour power overlooks the fact that means of production may themselves be living or naturally regenerative (e.g. animals, crops). Conceptualising economic production purely in industrial terms thereby overlooks this third option of living means of production. And these may also contribute to the production of value.

This argument has a certain plausibility, so that it must be confronted directly. In doing so, however, we must first grasp a particularly thorny conceptual problem that underpins the argument, namely the social/natural dualism. This distinction has been significantly criticised and deconstructed (e.g. Castree 2002, Harvey 1996, Macnaghten and Urry 1998, Murdoch 1997, Smith 2007, Whatmore and Thorne 1998). If, however, a strict distinction between nature and society, non-human and human, is untenable, then there are no longer any grounds for claiming, with Marx, that the latter alone can be the source of value, for the distinction itself has dissolved. Let us first, therefore, attend to this distinction. In doing so, I will argue, we must take two steps.

The first step is to concede many of the criticisms of a strict nature/social distinction and the presumption that this refers unproblematically to utterly distinct, abstract 'spheres' of reality. It is not the case, however, that Marx himself or (relational) Marxian analysis posits social/nature distinction as if these are 'real' categories. Rather, such analysis highlights how capitalist relations of production themselves have posited and strengthened such a distinction. This depends upon two intrinsic characteristics of capitalist accumulation: (1) the distinction of use-value versus (exchange-)value and the absolute priority of the latter renders the capitalist mode of production (CMP) blind to the former, except insofar as it happens to raise problems for seamless production of surplus value (Castree 2002); and (2) that the CMP itself needs to cement, and so *construct* in reality, the distinction between 'intra' and 'extra' economic processes, where the latter are taken (in their guise as *use*-values) as objective limits to production and so can be relied upon to present unbreachable barriers to production processes, hence allowing for 'objective' setting of the socially necessary labour time, i.e. value (and, where possible, in forms that are of maximal assistance to capital accumulation).

Regarding 'nature', then, point (1) indicates that, even though nature is a ubiquitous condition of and means of production in multiple (if not all) production processes, it is nevertheless, as a use-value, of no direct concern to *capitalist* production processes. The dominant understanding of 'nature' within a capitalist economy will thus be conditioned by the most useful conceptualisation of it from the perspective of (contemporaneous) capitalist accumulation, and not its own intrinsic dynamics or properties. Regarding (2), then, the 'objectivity' of the limits of production imposed by the use-value 'nature' was a valid or reliable presumption for a prolonged period in the history of capitalist development that enabled continued capital accumulation on the basis that 'nature' could be treated as a given and/or uncontrollable externality of the production process. The strict nature/social distinction thus served the purpose of making capitalist accumulation relatively stable, providing the grounds for objective measurement of the exchange value of commodities, upon which GCE, and so capitalist production, depends. It is thus the particular logic of capitalist accumulation that has imposed (at a particular historical moment) the strict natural/social binary (Bellamy Foster 2000, Castree 2002) *not* the presumption of this (relational Marxian) form of analysis. This presumption, however, is now increasingly being called into question, under the expanding logic of capitalism itself.

However, it is not merely that a Marxian analysis can accommodate a repudiation of the strict social/natural dualism. In fact, this process can only be made *intelligible* by such rejection and exploring the concrete, historical process of the social *production* of nature, or *socio-nature* (e.g. Castree 2002, Smith 2007). First, 'nature' as given non-human-produced phenomena has always been an essential pre-condition of all production processes, not just capitalist ones. All production processes work with and on such non-human natures and transform them, so that they are intrinsically socio-natures. Moreover, this hybridity seems particularly striking today (see below), so that the strict distinction increasingly seems manifestly untenable. This raises the question, however, of how this conceptual distinction is and has been imposed upon the messy, hybrid reality it claims (and seeks) to connote, not least when it seems to be such a poor approximation.

Second, therefore, a strict social/natural distinction is itself a social construction of particular socio-historical setting with its own genealogy. For reasons just discussed, the advent of capitalism as the dominant mode of production is at least a major conditioning factor in this history (as is the parallel development of the modern(ist) state, e.g. Scott 1998, Porter 1995). But as 'socio-nature', it also follows that this historical construction and maintenance of the social/natural distinction would also include the transformation of material life to conform (as much as possible) with or be amenable to description in these terms, i.e. the (attempted, incomplete, contradictory and messy) *transformation* of 'nature' (or rather, diverse nature*s*) in particular directions that fit with their conceptual capture in these terms. As Castree (2002: 138) puts it: 'an irreducible and varied world of natural entities are linked together under the qualitatively homogenous and one-eyed imperatives of the peculiar process Marx calls capital.' Given that actual efforts to impose the latter on the former occur always and only in specific circumstances, this also entails that one cannot actually study relationship of 'capitalism' and 'nature' as uniform abstract entities, but only as concrete processes of the interaction of actual capitalist economic processes and diverse and specific natures.

As such, adopting a broadly constructivist account of nature as 'socio-nature' (albeit a 'mild' or 'contextual' constructivism per Benton 2001) would seem to concede the argument to Brennan *et al.* But such a step of analysing 'nature' in this way does not and cannot licence the complete dissolution of this distinction, for at least two reasons: (1) the very idea of the hybridity of a socio-nature still presupposes (as a necessary condition of intelligibility) a distinction between phenomena that are produced and/or willed by humans and those that are not; and (2) even as the distinction is blurred and complicated in these necessary ways, the intelligibility of actual disputes about matters of (socio-)nature also presuppose a realism regarding the material, natural realities not constructed or construed by humans upon which a conflict arises. In both respects, therefore, to prohibit the social/natural distinction altogether is to paralyse not only critical analysis of its actual construction, form and impact in any given place and time – so that even avowedly constructivist work cannot avoid making such 'realist' assertions, even if these are only tacitly smuggled in – but also actual engagement in social life where it is mobilised continuously. As such, conceptual paralysis is more likely to be ignored than accepted, which in turn can only lead to inconsistency. A second step, proceeding full circle back to an abstract analytical dualism of social versus natural, must thus be made, but to a distinction that is now attenuated, fallible and fluid, and based on reasoned understanding (itself also fallible and subject to development) of real differences between the emergent properties of human and non-human natures *and* of the diverse social processes of construction of understanding and meaning regarding them (Benton 2001: 12 ff.).

Once the immanent necessity of maintaining an analytical distinction between social and natural is also acknowledged, however, we can also proceed to an abstract analysis that can conclusively show the impossibility, in principle, of non-human nature(s) as a source of value. In brief, value is only produced by labour because the possibility of getting more value from consumption of this (fictitious) commodity than was consumed in its production – hence the possibility of being 'exploited' in construction of (surplus) value – is a unique characteristic of (human) labour power. There are two elements to this: (1) the possibility of competition between means of production and so 'working' not merely 'being worked'; and (2), especially regarding *relative* surplus value upon which the exceptional productivity of capitalism has overwhelmingly depended, the possibilities of emergent productivity gains from cooperative work.[5] Both of these factors are dependent on a means of production that is both thinking and embodied, conditions that seem only to be met by human labour power.

Let us consider the example of a cow used in the production of milk. The cow is one of the 'means of production' of milk, along with its (possibly scientifically managed) feed, shed, the (more-or-less sophisticated) milking equipment and the farmhands' labour. But can the 'consumption' of the cow's use-value yield more exchange-value than was involved in its (re-)production, whether from generation to generation or day to day? When the use-value of a cow fails to find a buyer because it is simply not productive enough to meet current market standards, this situation can only be remedied by its owner (the farmer) offering it for sale at a lower exchange-value. Conversely, if labour power fails to find a buyer, its owner can likewise offer itself at a lower exchange-value but *also* can *compete* by offering to work for the same wage as its competitors but harder and/or longer. As such, the productivity of labour power is not a fixed and objective factor in the way that of a cow (or a Petri dish of bacteria or a field or, of course, a piece of machinery) is at any given time.

In short, labour power is unique in being able to *work* and not just *be worked*, and so being able to be *forced* to work more or less intensively. Indeed, even if we attribute higher levels of consciousness to other organisms than we currently do (and both the current extent and limitations of our knowledge demand that we must entertain this possibility, especially for social mammals), so that they are also capable of responding to pressure by forcing out more productivity, this is not to deny that *labour power* is the only source of value, but simply to extend this beyond *human* labour power to include other species with degrees of self-consciousness. Moreover, as regards the emergent productivity effects of collective work and the division of labour, such emergent effects are also dependent upon labour that is capable of working *together* and cooperating. Again, this presupposes certain cognitive and social capacities that (to our best current understanding) are not exhibited by other, non-human natures.

We may, therefore, contrast the *generativity* of 'nature' with (what may be alliteratively termed) the '*generosity*' of human labour, i.e. the ability or possibility to give of oneself for the production of surplus for others, albeit often largely a matter of coercion. Yet, this generosity itself arises on the basis of the unique position of labour power as both spontaneously (re)generative *and* amenable to the parasitism of calculation of rational efficacy, against 'nature' and machinery, which can respectively fit the first and second criterion only. For it is the fact that humanity can *actively* and *cooperatively* (or collectively) respond to the strictures of the latter, and thereby cram its own natural generativity into a rational cage, that is the mark of its 'generosity' of spirit.[6]

Moreover, this distinction between generativity and generosity depends upon the crucial Marxian distinction of use-value and (exchange-) value. For the former is concerned only and entirely with use-value, while it is only when subjected to the distinctly capitalist logic of (exchange-) *value* that the unique characteristic of human labour power that we have called its 'generosity' emerges. Labour power is thus unique as a source of value because it can simultaneously and in parallel produce both use-value and (possibly surplus) value. Yet, this distinction, in turn, allows us to situate both the prima facie plausibility and the error of the 'nature as source of value' argument. A representative example is Brennan's leap from Marx's analysis of the capitalist labour process under socio-historically specific conditions of GCE to the trans- (or even a-) historical 'energetic' analysis of the production of '(surplus) value'.[7] On the one hand, this seems to challenge an unwarranted anthropocentrism to the Marxist analysis, but, on the other, *only* on the basis of a straightforward conflation of levels of analysis, shifting from the examination of exchange-value to use-value.[8]

Finally, as discussed in more detail below, the capitalism-mediated construction and cementing of the natural/social distinction in a different era or techno-economic industrial paradigm to the present is now subject to assault from the very expansion of capitalist relations of production. For while treating nature as an objective limitation on capitalist production has afforded the social stabilisation of market expectations that are presupposed by systemic markets, leaving natural processes, on which so many production processes depend, outside the control of the capitalist also brings considerable costs. As Smith (2007: 11) puts it, 'dependence on the availability of external nature for every cycle of production represents a considerable obstacle and source of insecurity for capital'. With the advent of biotechnological techniques in particular, these externalities, beyond the ken of the capitalist for much of history, are increasingly subject to attempts to bring them under such control. We close here, however, simply by noting that making sense of this change again demands analysis in terms of socio-nature.

2.4 Knowledge in a capitalist economy

We now turn to consider whether or not knowledge may be a source of value. Given what we have already discussed regarding nature versus labour, there is no need to rehearse further the argument that knowledge as a 'product', i.e. a body of learning, cannot be a source of value, even to the extent that it is itself a 'means of production' (including of further knowledge). However, knowledge arguably refers not just to the learning itself but to the process of its production. In short, the pertinent question is 'cannot knowledge *labour* be a source of value (and perhaps one of increasing importance)?'

To answer this question, we must start by considering the place of knowledge in the totality of the capitalist economy (before turning, in the next section, to consider whether knowledge can be a source of value). As Figure 2.1 illustrates, even just considering knowledge as a third input in a production process (alongside the traditional economic ones of capital and labour), and the processes of *their* production and circulation in turn, leads swiftly to an exceedingly complicated picture. 'Knowledge' enters into economic activity at multiple points: not just at the level of know-how relevant to a particular production process, but also (inter alia) in the (continuous) training of labour, the transformation through innovation of the firm itself, the management and maintenance of circulation (via logistics, marketing and financial decision-making) and multiple policy decisions. The potential reflexivity of knowledge, applying knowledge to itself and its own further production, adds yet further complexity. At the very least, therefore, this schematic diagram thus once again reminds us that there is no single 'economics of science' (or 'of knowledge') that can be exhaustively analysed in the abstract alone.

Moreover, Figure 2.1 also serves as a prompt that, in order to understand the role of knowledge in the economy, we must also understand the economy of the production of knowledge (including scientific knowledge) itself. This leads us directly to engagement with the conventional 'economics of knowledge' (Foray 2004) and/or 'of science' (David and Dasgupta 1994).[9] As discussed in Chapter 1, this conceives of knowledge as a non-rival and non-excludable economic 'public good' so that there is insufficient incentive to produce it at socially optimal levels on the basis of market-based private appropriation. The resulting market failure thus justifies public subsidy to ensure appropriate levels of production. Moreover, regarding knowledge, the market failure is argued to be especially acute given a number of characteristics of (scientific) knowledge as:

- cumulative and interactive in its production, viz. 'standing on the shoulders of giants';
- extreme in its indivisibility and durability; and
- subject to problems of asymmetry of information regarding the particular worth of any given piece of knowledge, complicating standard arm's length exchange in that a buyer cannot 'test' the product before buying it, without thereby irreversibly acquiring it.

In all cases, therefore, (scientific) knowledge is conceptualised as 'general' in the sense that it is universally applicable, and hence can be used with equal efficacy anywhere and by anyone.

In fact, this acute market failure in the production of knowledge need not always lead to negative outcomes. Indeed, where knowledge may be produced without significant economic constraints, these characteristics may also lead to 'positive

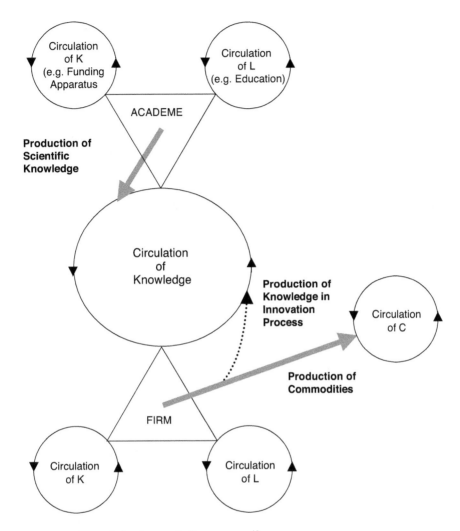

FIGURE 2.1 Knowledge in a capitalist economy[10]

externalities' that generate not the conventional 'tragedy of the commons', where free-riding undermines all such activity (e.g. Hardin 1968), but to a 'comedy of knowledge commons' (Foray 2004), where qualitative leaps in knowledge-based social progress emerge through the spontaneous and open activity of networks of multiple, dispersed knowledge agents.[11]

There is clearly much that is intuitively plausible about these arguments, especially with the evidence of both significant and growing public funding of science on the one hand, regarding the 'negative externalities' of knowledge production, and the emergence of just such knowledge-producing networks on the other, regarding the 'positive externalities'. However, this argument is also, at best, half the picture and hence harmfully incomplete. In particular, we must ask 'what is meant by "knowledge" here?' As many commentators have noted (e.g. Callon 2002, Mirowski and Sent 2002a), these arguments employ a limited understanding of knowledge as information, hence a naturalised thing or commodity that is little different to widgets.

Moreover, this is even true of the 'new' economics of science (David and Dasgupta 1994), which is expressly founded on challenging the limited definition of knowledge applied in the first generation of such arguments. The NES adds the distinction between codified and tacit knowledge, arguing that the 'public good' argument applies only in the former case, while both are equally important. Yet tacit knowledge (including science, following Collins 1974, 1985) is embodied in particular people and so is rivalrous, appropriable/ appropriated and, as Foray (2004: 17) puts it, 'fragmented, partially localized and weakly persistent'. Nevertheless, it continues to employ a notion of 'knowledge' that, whether codified or tacit alike, is amenable to an exhaustive analysis in purely abstract terms, hence still treating it as a 'thing'.

Among the most important critiques of this approach is that of Callon (1994, 1998a, 1998b, 2002). For the new ecconomics of science (NES), the central assumption that affords the matching of the distinctions of public versus private good, with codified versus tacit knowledge respectively, is that the former dualism thus describes *intrinsic* features of the relevant knowledge. Conversely, for Callon, the public good properties of particular knowledges emerge as *extrinsic* network properties under particular conditions that themselves must be constructed and hence economically accounted for, e.g. regarding acquisition, set-up and/or transaction costs (Callon 1994).[12]

Instead of knowledge as public versus private good, therefore, Callon substitutes the distinction between 'emergent' and 'consolidated' configurations of knowledge. As Table 2.1 summarises, emergent configurations are those in which (scientific) knowledge is novel and searching to construct the social and material relations needed to stabilise it as accepted as 'fact', while consolidated configurations have already achieved this (for the time being). This approach also then builds upon insights from science and technology studies (STS) (especially those associated with actor-network theory (ANT)) that the construction of the stabilised social phenomenon of sedimented scientific knowledge is inseparable from the (co-)production of an equally stabilised socio-economic order. While this perspective, like NES, thus presents two ideotypical cases, its emphasis on concrete processes of socio-technical construction leads to a quite different programme of research, focusing on empirical analysis of the multiple hybrid forms that actually exist in given times and spaces.

While Callon's development of the 'public good' argument is undoubtedly an improvement on NES, though, there are at least two major shortcomings remaining in his analytic framework. First, the wholesale shift from exclusively intrinsic to exclusively extrinsic characterisation of the economic characteristics of particular knowledges is unnecessary and unhelpful. In short, while empirical attention to

TABLE 2.1 'The main features of emergent and consolidated networks'

	Emergent configurations	*Consolidated configurations*
Knowledge	• Statements + instruments + embodied skills	• Statements are information because embodied competences are duplicated
	• Non-substitutability between codified and embodied knowledge	• Codified and embodied knowledge are relatively substitutable
	• Knowledge is private: rivalry and excludability	• Knowledge is public – i.e. non-rival, non-exclusive – in the networks in which it circulates
	• Knowledge replication = laboratory replication	• Knowledge replication = coding and replicating strings of symbols
	• Local knowledge is generalised through successive and costly translations/ transportation	• The degree of universality of knowledge is measured by the length of networks
States of the world	• List and identity of social and natural entities in constant reconfiguration	• List and identity of social and natural entities are known
	• States of the world revealed, *ex post*, through trials and interactions	• All states of the world are known *ex ante* and the probability of their occurrence can be calculated
	• Uncertain and vague knowledge uses (this depends on the scope and state of the translations; they become standardised and stabilised along with the networks)	• Uses of knowledge are predictable; they are more soundly established when there are multiple connections (principle of network externalities)
Modalities of action	• Programmes only exist *ex post*, as the outcome of action	• Research and innovation programmes (list of problems to solve and operations to accomplish to reach a solution) are defined *ex ante* and provide a framework for action (coordination)
	• Mutual learning	• Rational expectations
	• Cooperation as an obligatory passage point for action, i.e. for translating identities and interests and for negotiating the content of knowledge	• Cooperation is a strategy for cost- and risk-sharing or for consolidation of power positions

Source: Callon (2002: 298).

concrete particulars is undoubtedly necessary, knowledge per se remains a real, emergent social (i.e. socio-material) phenomenon with real characteristics that are amenable to informative abstract analysis, even while this cannot exhaust or fully determine understanding. The characteristics of knowledge(s) are not, therefore, *just* extrinsic, and so general or tendential problems associated with the economics of *knowledge* in particular are indeed identifiable; including its different and limited possibilities for capitalistic production.

For instance, both non-rivalry and the generality upon which it is based are indeed contingent achievements of particular knowledges/claims, as Callon argues. But there is also an intrinsic dynamism towards (realisation of the *potential*) universality of knowledge as a necessary condition of its intelligibility as 'knowledge', even if this universality is thereby a contingent achievement and not an immediate or unmediated actuality. And this, in turn, mobilises an immanent drive to construct the networks for the (contingent) realisation of generality and non-rivalry, in order to maximise productivity of that knowledge, hence explaining why emergent configurations tend towards consolidated ones. Similarly, it is an abstract, but no less real, characteristic of knowledge that its production is exceptionally hard to distinguish from its circulation, reproduction and consumption (e.g. Hicks 1995), making application of conventional categories of economic analysis problematic.[13]

Second, Callon's analysis, while reframing the 'public' versus 'private' good distinction, accepts it as a valid starting place for economic analysis of knowledge production. Yet, just as the socio-material conditions for stabilisation of (scientific) knowledge must be constructed, the same is also true of the 'market' as the pre-eminent social organisation for coordinating economic activity. The 'market' for knowledge must thus *itself* be constructed, as there is no such thing as a pre-existing 'market' of knowledge. It follows that the basic presumption of the 'market failure' argument is unfounded, and Callon's challenge does not go far enough. Instead, following the earlier argument above regarding the CMP, our approach is to explore what must be the case for knowledge to circulate and *be produced* under capitalist conditions, i.e. as a 'fictitious commodity'.

In doing so, Callon's two-fold schema of consolidated and emergent configurations provides an informative basis given the advance it represents upon the NES. However, in order to take into account the criticisms just discussed, we translate his schema into questions of the economic problems associated with knowledge as fictitious commodity and markets for knowledge labour power (KLP) respectively. The former concerns knowledge that can and has been codified and so can be bought and sold, and so relates to the consolidated configurations and public good-like knowledge of Callon's typology. Conversely, the problems of markets for KLP will be shown to relate closely to the issues facing emergent configurations and knowledge as a private good. As this analytic framework involves abstraction, it also, perforce, cannot avoid certain simplification, but by focusing on factors that are emergent realities, it allows us to bring out important and real social dynamics (Sayer 1992: Chapter 3, Lawson 1997: Chapter 16). Hence, by reframing Callon's distinction – itself a reframing – in this way, we may approach closer to a faithful understanding of the economics of science and knowledge production. We can also, then, return to the task at hand, namely assessing to what extent 'knowledge' is a source of value.

2.5 Knowledge as source of value?

We have already concluded that for knowledge to be a source of value, it is as KLP, i.e. the latter of the two issues just discussed. We cannot focus exclusively on this issue, bypassing consideration of the economic problems of knowledge as fictitious commodity, however, because there is close interaction of the two phenomena, such that they must be examined together:

(a) First, the construction and stabilisation of viable markets for KLP – hence as one (let alone the sole or major) source of value – is hugely and intrinsically problematic. But so too is the production of knowledge as a fictitious commodity, without jeopardising future production of knowledge.

(b) As such, neither the emergent nor consolidated configurations, associated with issues of KLP markets and knowledge as fictitious commodity respectively, are stable in themselves. In search of stable profitability or competitive advantage, therefore, there is an intrinsic drive between these two poles. It follows that the two must be studied alongside each other to understand the emergent, systemic dynamic of a capitalist knowledge economy.

Regarding (a), first, the problems of knowledge as codified and sellable fictitious commodity, within consolidated configurations, are, at least on first sight, familiar. Phrased in terms of 'public good', they raise the problems of making a profit given the difficulties of private appropriation. To this analysis, Callon adds (or substitutes) the need to construct socio-technical networks that in turn construct knowledge that is sufficiently general and with marginal costs of production/transmission low enough for this problem (if, indeed, it is a 'problem') to emerge. However, he overlooks the need to construct not just the knowledge but also the *market* for knowledge. The core of these problems of knowledge as fictitious commodity is thus best understood in terms of the problems of sustaining an exchange or market-based economy underpinning knowledge production in which the socially necessary labour time of production (and hence its value) does not equal (or tend asymptotically to) zero, as is the case where it may be almost freely and instantaneously copied and distributed. In this latter case, knowledge can only be produced in the context of a gift economy.

This terminology thus draws out not only Callon's focus on the need to construct (scientific) knowledge *as* (general, universal) knowledge per se but also the need to construct knowledge as *fictitious commodity*, this also always being a contingent and socio-historically particular achievement. The relation of consolidated configurations of knowledge to value is also thereby elucidated: knowledge must be valorised, given an exchange-value, in order to be produced in a capitalist production process. Yet, this is both primarily a problem for *capitalist economies* that cannot simply be presumed as 'the natural order of things', and hence not necessarily a problem for knowledge per se; and insofar as commodification of knowledge is achieved, it immediately raises further problems either as regards incentives for its (continuing or prior) production or, insofar as it is appropriated, for its future development.

Whereas the economic problem for consolidated configurations is one of sustaining viable markets for commodified knowledge, for KLP, in contrast, problems arise regarding the creation of a market price for this commodity *in the first place*. These problems may be understood in terms of those associated with emergent configurations and tacit knowledge, embodied in particular knowledge workers (i.e. people). Again, for Callon, the emergent configuration is one in which new findings are striving to construct the socio-material networks that will stabilise their social status as (potentially economically valuable) 'knowledge'. At this stage, therefore, the knowledge is largely inaccessible and embrained and embodied in its particular progenitors.

This account, however, again overlooks the importance of construction of a *market* for this KLP. And this demands that a (systemic) exchange-value, representing the socially necessary labour time incurred in its (re)production, can be accorded to this KLP. For such a market price, or exchange-value, is essential if KLP is to contribute to value creation, since surplus value can only be created if labour power can itself by purchased at exchange-value, i.e. if labour has itself already been commodified. The construction of knowledge *labour power* as fictitious commodity thus, inter alia, places an imperative upon the development of the socio-technical means (or 'technologies', in the Foucauldian sense) to be able credibly to measure or quantify (what may be socially accepted as) 'objectively' the exchange value of a particular KLP.

Consideration of the general characteristics of *knowledge* labour power, however, reveals three challenges for such market quantification that are likely to be particularly acute in this case (though not entirely absent for other forms of labour), namely:

1   Quantification of the concrete contribution of a particular KLP to a production process: while it is relatively straightforward to assess that a worker has completed 500 steps in time t in the production process for a car, say, a similar calculation for knowledge work, and especially that associated with individual, tacit knowledge, is much harder to assess. How, then, does the firm discipline its labour force (not least, with the threat of dismissal) to meet required targets, thereby maximising the production of (probably relative) surplus value, on the basis of their actual measurable contribution to output? If surplus value cannot be reasonably assured, however, the worker will not be hired.

2   Quantification of the qualifications or capability of a particular KLP: a knowledge worker is hired at a price calculated on the basis of what he/she can contribute to the improvement of productivity in comparison with the

value needed for the (re)production of that labour power, where the former must exceed the latter to be profitable. Yet, especially given problem 1, which deprives the capitalist employer of the means of definitive test, it is extremely difficult to quantify the *potential* contribution of a particular KLP. How, then, does the capitalist work out whether it is paying too much for the knowledge worker?

3   Knowledge labour power markets: finally, both of these problems are particularly marked for exceptional KLP, associated with emergent configurations and tacit, embodied knowledge. In all cases, where labour markets function reasonably smoothly, this is because the labour power on sale is in dynamic market competition with n other workers, who can also complete x processes in time t, or if not, be tested and replaced (via 1 and 2 above). Yet, as the number of workers claiming and/or believed to have such capability decreases, the conditions of perfectly competitive markets also dissipate. As many forms of knowledge labour involve relatively highly qualified and/or unusual capabilities, markets for them may tend to these monopolistic or oligopolistic conditions.

These problems thus impel the construction of the socio-material means by which KLP may be thus quantified and subjected to market competition. Hence, much knowledge work can and has been commodified, in fact, on the back of, and creating demand for, a lot of further (profitable) knowledge work expended in construction of these technologies (e.g. 'knowledge management').[14] Successful commodification of knowledge labour, however, transforms KLP from the exceptional capability of particular individuals to mass employment of routinised and low(er)-skilled labour. To be sure, this may well remain profitable for a capitalist firm for a significant period, but the very process of creating the conditions to subject KLP to competition also thereby opens up the firm itself to competition and in forms of economic activity that may have particularly low barriers to entry *as* routinised knowledge work.

This, therefore, takes us to (b). For it is not simply that both consolidated and emergent configurations are themselves unstable, reflecting the intractable problems of knowledge and KLP respectively as fictitious commodities. But the apparent, albeit only ever temporary, solution of the problem at each pole is, in effect, to strive to move towards the other, for at least two interrelated reasons, regarding both the capitalist search for profit and the cumulative and interactive productivity of knowledge production per se, i.e. for reasons of exchange-value and use-value respectively.[15]

Regarding the former, in a consolidated regime, a capitalistic producer of knowledge has succeeded in constructing the socio-material conditions needed for there to be widespread demand for its knowledge product. Yet, this will have simultaneously created commensurate problems regarding the appropriability of its knowledge product. Thus, in seeking to solve problems of appropriability associated with the resulting 'public good' characteristics of knowledge as fictitious commodity, a capitalist producer must innovate and/or seek to increase the stickiness (including by the temporary legal monopoly privileges of intellectual property rights; see Chapter 4) and tacitness of the knowledge upon which the consumption of the codified, commodified knowledge depends. This drives the producer towards the emergent configuration pole, and especially towards the search for appropriation of KLP that is sufficiently exceptional that it will produce knowledge that can generate monopoly super-rents. Yet, in doing so, the associated problems of emergent configurations emerge instead, regarding the difficulty of valuing the exceptional KLP that underpins the production processes capable of generating innovation super-rents. Completing the picture, attempts to resolve these problems in turn lead to attempts to construct networks that evince sufficient demand without undermining the super-rents through disclosure of the knowledge product, i.e. passing back to the problems of the consolidated configuration.

A similar dynamic can also be seen in terms of the use-value of maximising the productivity of knowledge production, or rather, at least avoiding destruction of such productivity, thereby sweeping the rug from beneath one's feet, in the medium term if not immediately. Exchange-value is dependent upon use-value, both in that money will not be paid for a commodity unless it (is believed to) serve(s) certain purposes of the purchaser and, in the context of production processes, (surplus) value cannot thus be produced without the concrete, socio-material pro-duction process of the commodity. Accordingly, capitalist production must take the constraints of a production process (at the level of use-value) into account, even though it may not do so in the short-term where the demands of maximising exchange-value conflict with such constraints, i.e. given that (as discussed above regarding nature) capital is blind to use-value.

As regards knowledge production, however, the key to any such maximisation of the (use-value-level) productivity of knowledge is its cumulative and interactive nature. Yet, this demands in both cases, whether regarding knowledge as fictitious commodity or KLP markets, that capitalistic organisation of knowledge production be significantly supplemented and encompassed in systems of *non*-capitalist knowledge production, including spontaneous knowledge-sharing communities and publicly funded research.

In the case of consolidated configurations, where a successfully stabilised fictitious commodity of knowledge has public good properties, to the extent that appropri-ation of such knowledge is successful, it also imposes constraints on the circulation and cumulative fecundity of such knowledge, most obviously in the 'deadweight' monopoly burden of intellectual property rights (IPRs, or more accurately, intellectual monopoly privileges (IMPs); see Drahos and Braithwaite 2002). This must therefore be 'balanced' by rules of public disclosure to encourage the dynamic social productivity of knowledge production, building on the commodified knowledge. Conversely, for KLP markets in emerging configurations, the need is to construct socio-material conditions and networks that will underpin market demand. The capitalist knowledge producer wants to maximise the productivity of that knowledge in order to make it attractive to as many consumers as possible, thereby maximising demand, while also simultaneously cementing competitive position in knowledge networks for future innovation and associated rents. But this depends upon making multiple connections with others who can suggest and envision uses of this knowledge and put them to use – uses that the original source of the knowledge often cannot conceive, let alone implement – and the problems of asymmetry demand that these relations cannot be arm's length market transactions.

Thus, while the extent of the novelty of the emergent knowledge is proportional to the potential super-rents of innovation – the innovator (or rather the *owner* of the innovation) clearly seeking for these to be as large as possible – it is also proportional to the level of new connections needed in order to build upon the innovative knowledge's productivity before the 'potential' use-value (and hence exchange-value via market demand, maximising profit in turn) of the knowledge is appreciated and thus potentially realised. Clearly, network positive feedback loops and scale effects are relevant here, as when a fax becomes more useful, and hence more valuable (at least until competition drives the price *down*), the more *other* people also have one.

In limine, therefore, the former problem can lead to a 'tragedy of the anti-commons' (Heller and Eisenberg 1998) or blockage, in which productivity of knowledge is stymied by too many private knowledge claims (such as a patent thicket) excluding the productive, interactive and cumulative development of further knowledge; while, in the latter case, refusal to compromise and share a unique knowledge-based competitive advantage will simply result in it having no economic impact whatsoever. The emergence in recent years of the networked firm and (possibly global) innovation networks, inter-firm strategic alliances or aggregation of 'knowledge campuses' for particular projects, and the informal waivers of rights (perhaps trading licences to IMPs as bargaining chips) – all of which are especially prevalent in knowledge-intensive, hi-tech industries – are perfect examples of the multiple concrete attempts to straddle these tensions.

2.6 Conclusion: Labour (power) as source of value

We have examined the economics of nature and knowledge and their potential to act as sources of economic value within capitalist production. Against a central claim of the contemporary policy discourse of a 'knowledge-based bio-economy', neither nature nor knowledge per se can act in this capacity, though knowledge labour (or, more accurately, knowledge labour power) can. Even regarding the latter, however, capitalistic markets of labour power are highly and intrinsically unstable and in need of constant non-capitalistic or extra-economic support.

If neither knowledge nor nature are (self-sustaining) sources of value, however, the KBBE *cannot be* a fully capitalist production of knowledge, against boosters and jeremiads alike. Instead, it becomes apparent that the associated problems of subjugating diverse natures, knowledges and knowledge-intensive human capacities to the one-dimensional logic of capitalist value are insoluble in the abstract, and hence 'solved' always only temporarily and in concrete. This depends on politics (e.g. subsidies, public-platforms, (de)regulation, IPR laws, etc.) and new organisational innovations or socio-material technologies of measurement that make possible the quantification necessary for capitalist exploitation of these phenomena (itself a key role of scientific work), so that we may talk of the *construction* of value in its strict, theoretical sense. These diverse mechanisms also then illustrate the central role of – and hence, open up and demand the empirical investigation of – the diverse forms of socio-political agency in the 'regularisation' of a KBBE, particularly regarding establishment of various spatio-temporal fixes that stabilise the possibility of continuing capitalist accumulation and externalise its costs to other people in other times and places.

Our investigation of the KBBE cannot end here, however, but raises further questions. In particular, if the KBBE is not as described by these dominant analyses, an alternative account is needed. In constructing this, however, we must pay due heed to the causal efficacy of the KBBE policy discourse as well as the associated conjunction of political economic structural conditioning, which renders some interpretations of reality more powerful than others, in order to explain what is actually being built as a 'knowledge-based bio-economy'. But we have also seen that abstract analysis regarding the political economy of a KBBE, and its intrinsic contradictions, must be supplemented by and developed alongside analysis of concrete actuality. We therefore will develop the argument with such empirical analysis of a sector that is at the heart of KBBE agenda: agriculture.

Further reading

Callon, M. (2002) 'From Science as an Economic Activity to Socioeconomics of Scientific Research: The Dynamics of Emergent and Consolidated Techno-economic Networks', in P. Mirowski and E.-M. Sent (eds), *Science Bought and Sold*, Chicago: University of Chicago Press.

Castree, N. (2002) 'False Antithesis? Marxism, Nature and Actor-Networks', *Antipode* 34(1): 111–146.

European Commission (2010) *Europe 2020: A Strategy for Smart, Sustainable and Inclusive Growth*, Brussels: EC.

Powell, W. and K. Snellman (2004) 'The Knowledge Economy', *Annual Review of Sociology* 30: 199–220.

3

THE KBBE REALITY

The case of agriculture

3.1 Introduction

In Chapter 2, we examined one of the central claims of discourses regarding the knowledge-based bio-economy (KBBE), regarding nature and knowledge as sources of value, and found it wanting. In this chapter we set out an alternative conception of the KBBE as it really exists. In doing so, we will address four issues: the relations among financialisation, neoliberalism and KBBE discourse; that, as a result, KBBE discourse is not just empty words but (partly) constitutive of the reality it claims only to describe; a characterisation of the real KBBE that is being constructed under financialised neoliberalism in the crucial domain of agriculture, in terms of a four-dimensional attempt at its real subsumption by capital; and, finally, against the second central claim of KBBE discourse, the vacuity, if not straightforward falsehood, of its claims of ecological sustainability.

3.2 The real KBBE – Financialisation, neoliberalism and globalisation

Given the inadequacy of KBBE discourse to its subject matter, its fundamental confusion and wrong-headedness, the temptation may well be simply to dismiss it. Yet it remains a policy discourse of enormous power, especially in the EU, where it is explicitly shaping both economic policy and scientific funding. We must therefore go beyond analysis of its substance as a discourse and examine it also as a sociological phenomenon demanding explanation, especially regarding its emergence and power.[1] In this section, we will examine the interrelations among KBBE discourse and three particularly striking features of the global(ising) political economic context from which it has emerged, namely financialisation, neoliberalism and globalisation.

First, what is financialisation? Consider the current phase of financialisation as an illustration. This has arisen out of the widely observed crisis of 'Fordism', or what may be more accurately defined as industrial, mass-production capitalism in conjunction with a political economic settlement at nation-state level that instituted a relatively stable class compromise between industrial capital and labour in the form of the Keynesian welfare state (Jessop 2002a). Given the inherent contradictions of capital accumulation, such political economic settlements are always needed to manage, or rather 'regularise', capitalist economic growth and make it relatively stable and predictable, if only ever temporarily. As the limits of this arrangement became evident from the late 1960s and through the 1970s in the emergence of stagflation across the developed economies of the global North, investment in finance rather than in productive industrial capital became progressively more attractive, such that financial-sector profits had risen to 39 per cent of total US corporate profits in 2001, as against 14 per cent in 1981 (Brenner 2004: 76, Strange 1986, 1998).[2]

Perhaps the most striking example of this is the industry associated with the eponym of Fordism itself, the car industry of the US. For instance, Blackburn (2006: 44) notes:

> By 2003, 42 per cent of the group's profits were generated by GE Capital. In the same year GM and Ford registered nearly all their profit from consumer leasing arrangements, with sales revenue barely breaking even. . . . In 2004, the General Motors Acceptance Corporation (GMAC) division earned $2.9 billion, contributing about 80 per cent of GM total income.[3]

This 'financialisation' of the economy, however, is not just a matter of industrial giants seeking their profits in finance instead of their 'core' productive businesses. Rather, it is marked by the domination of the economy (both in quantitative/ economic and *political* terms) by the workings of finance making money from money. This singular fact raises serious questions about the emergence, let alone current existence, of the KBBE, for it is apparent that it is *finance*, and not knowledge or 'nature' (viz. biological or ecological technologies), that is the dominant sector of the economy – even following the great financial crash of 2008 (e.g. Stephens 2010; see also Hacker and Pierson 2011, Shaxson 2011).[4]

In short, therefore, financialisation connotes the economic and *political* dominance of finance capital that emerges in the context of a systemic accumulation crisis and the associated crisis of regularisation regime. As such, financialisation involves not merely a progressive shift of investment from industry to finance, but (especially in the core economies of the global capitalist economy) a political counter-revolution that transforms the calculations of government to prioritise the demands of finance capital over those of both productive capital and, of course, labour. Financialisation is thus associated with a tide of radically pro-capital(ist) politics that, in turn, radicalises the state (at both internal/national and external/international levels) through effective capture by a 'historic bloc' (Arrighi 1994, Gramsci 1971) centred around finance capital.

One particularly important result is that the distinction of 'public' versus 'private' sectors, which is always problematic (inadequate, polysemous etc.),[5] becomes particularly unhelpful under conditions of financialisation, especially if it is deployed in terms of a Manichean political struggle between 'the public' and 'the private' sectors. Rather, where the needs of this historic bloc depend upon the mobilisation of the full power of the state's apparatus, it will often be able to compel the state to act effectively as an arm of the 'private sector'. Two classic examples of this process are the bank bail outs of 2008 (with an ex-Goldman Sachs US Treasury Secretary bailing out some of Wall Street, including his old firm, but letting the competition, Lehman Brothers, collapse) and the Trade Related Intellectual Property Agreement (TRIPS) for global intellectual property rights (see Chapter 4). More generally, the crucial importance of 'extra-economic' regularisation of capitalist accumulation by the 'state' combined with the dramatic changes to the political economy driven by financialisation leads to processes of de- and re-regulation of economic activity, that redraws the very boundaries of the state itself.[6]

This leads us directly to our second issue, neoliberalism, the political project associated with the counter-revolution underpinned by financialisation (Birch and Mykhnenko 2010, Castree 2006, 2008, Peck and Tickell 2002). Neoliberalism is both a political project to restructure the global economy (leading to processes of 'neoliberalisation') and a political ideology providing a unifying vision for disparate political interests – hence a central aspect of the historic bloc of financial capital that makes it 'historic' in this sense. Classical (economic) liberalism is the political creed that argues for the socially optimal allocative efficiency and non-zero sum gains of laissez-faire free-market capitalism. It is thus intrinsically inimical to state involvement in the economy. Conversely, based on the seminal works of Friedrich Hayek and the Mont Pélérin Society in the 1940s (Miller 2010, Mirowski and Plehwe 2009), *neoliberalism* is new ('neo') in the sense that it is a radicalised liberalism in which the programme of liberalisation of economic activity must be, if necessary, driven through, against political resistance, by the state; hence forcing economic activity to be 'free'.[7] It is thus a curious hybrid of capitalist free-market ideology and state power, the 'mobilization of state power in the contradictory extension and reproduction of market(-like) rule' (Tickell and Peck 2003: 166). We see, therefore, how the state has a key role in regularisation of capital accumulation not only within established cultural political settlements (as per the Keynesian welfare state) but also in their *expansion* and *construction*.

Yet, there is no such thing as 'neoliberalism' per se, as a purely abstract social force in the world. Rather, neoliberalism manifests only in diverse state-led projects of de- and re-regulation, privatisation, liberalisation and marketisation that attempt to resolve (or at least defer or displace on others) the economic crisis of Fordist accumulation. Such transformation of the political economy and associated policy thus tends to support the further development of two (mutually reinforcing but also not entirely compatible) political economic trends that (promise to) move beyond Fordist industrial capitalism: financialisation (as discussed above); and attempts to *construct* a 'new economy', built on the opening of seemingly huge new opportunities for non-zero sum, radical technological innovation and growth of productive capital.

With promise and expectation playing such an important role in directing these processes of neoliberalisation, there is clearly a need for diverse policy work, discourses and frameworks. Evidently, with its vision of a new economy of sustainable knowledge- and bio-based industries, KBBE discourse is one such example, though it is merely the latest incarnation of multiple policies since the 1970s, regarding the 'post-industrial economy' (Bell 1974), 'learning society' (Archibugi and Lundvall 2002, Lundvall 1996), 'knowledge-based economy' (OECD 1996) and 'information economy' (Castells 1996).

Similarly, the particular focus on current efforts to build a new economy is, of course, that summarised by KBBE discourse. Yet, this results in various policies and policy-supported projects that are attempting the neoliberalisation of 'knowledge' and 'nature'. Or rather, just as neoliberalism is in fact a diverse, iterative, shifting and contradictory set of concrete political projects, so too, regarding the KBBE, it is characterised by the neoliberalisation of diverse knowledge*s* and (socio-)nature*s*; for example, 'privatization (e.g., of land), marketization (e.g., of air), deregulation (e.g., of environmental protection), reregulation (e.g., biodiversity), liberalization (e.g., of trade in resources), competitiveness (e.g., in resource markets)' (Birch *et al.* 2010: 2900). In short, the concrete processes of the actual construction of a 'KBBE' involve multiple dimensions, so that again we see the need to study responses to these trends in concrete. Accordingly, we will explore the example of the neoliberalisation of agriculture below by way of illustration, and as a crucial socio-economic sector in its own right.

Before we turn to this, however, let us note three further points. First, while comprehensive analysis of the actual KBBE demands attention to empirical detail (whether at 'micro', 'macro' or 'meso' level), an abstract analytical framework is necessary to understand such data. Using a relational Marxist approach, we may think of the political economic task of such 'new economy' policies in terms of the construction of the sociomaterial conditions that engender new opportunities for the production of (relative) surplus value, in the form of new forms of productive labour for capital (PLK). Recall that PLK is labour power expended in conditions of (Fine 1986, Mohun 2003, Savran and Tonak 2002):

1   waged labour;
2   employed by capital (i.e. with a view to making a profit); and
3   in the sphere of production (of commodities, whether products or services).

There are multiple possible concrete ways in which new PLK can be generated at any given time and we will consider many such examples below. In principle, however, there are three possible avenues for the introduction of new forms of PLK, namely:

1   to transform existing labour practices that are currently in the sphere of production but non-productive for capital into productive labour;
2   to introduce new opportunities for capitalistic exploitation of productive labour; or
3   to increase the productivity of existing productive labour by increasing the intensity of relative surplus value.

Point 3, in turn, may also take at least three forms, namely regarding the decrease of either:

(a)   the socially necessary labour time (SNLT) of a productive labour process;
(b)   the socially necessary turnover time (SNTT) of the commodity (Harvey 1982); or
(c)   the naturally necessary production time (NNPT) of a labour process.

All three of these involve innovative transformation of the means of production, thereby affording the increased intensity in the exploitation of labour power and the commensurate possibility of super-profits to the successful entrepreneur. We will bring out examples of these various mechanisms below.

Second, we must also note (at least some of) the intimate connections amongst the three processes – of neoliberalism, financialisation and KBBE (discourse) – and the current globalisation of the capitalist economy. For contemporary 'globalisation' is a process marked by transformation of global economic regulation according to neoliberal prescriptions, with significant liberalisation of global trade (though with notable and important exceptions, such as the EU's Common Agricultural Policy) and the growth of transnational corporations, as well as characterised by equivalent contradictions to those of neoliberal capitalism, as contradictory processes of de- versus re-regulation, regarding the latter, parallel pressures to de- and re-territorialise production processes regarding the former (Jessop 1999). Similarly, it is in financial markets in particular, as the most (easily) deterritorialised (as the least materialised) of economic sectors, that globalisation has proceeded first and farthest, leading to mutually constructive feedback loops that deepen the global political economic dominance of Organisation for Economic Cooperation and Development (OECD)-country-based finance capital (particularly Wall Street and the City of London). Finally, regarding the KBBE, there are multiple important connections between globalisation and knowledge-intensive and/or bio-based industries. A non-exhaustive list of these relations includes: the growing importance of global(ising) networks of production and of innovation in pursuit of competitive advantage, particularly in knowledge-intensive sectors (Archibugi and Iammarino 2002, Ernst 2008); the search for global markets, again particularly in such industries, given the growing (globalising) competition and rising costs of research and development (R&D) (Archibugi and Iammarino 2002); and the exceptional global political power and profitability of major knowledge KBBE transnational corporations (TNCs) (Dicken 2007). In short, there is an irreducible global dimension to the neoliberal construction of a KBBE.

3.3 Financialisation and the KBBE

Finally, we have yet to complete the circle of our analysis by connecting the KBBE and financialisation. It needs little imagination to see that there is a striking fit between the current (and doggedly persistent, even after the upheaval of the 2008 crash) dominance of finance capital, its drives to restructure the economy in search of new sources of surplus value and the fundamentally false, permanently deferred ('jam tomorrow . . .') and hugely hyped promissory tenor of much KBBE discourse (Birch 2006). There is thus a clear interactivity or mutual constitution between the KBBE discourse and financialisation.[8] This takes effect across numerous dimensions. Let us consider five of these (in no particular order) regarding the dependence of the KBBE discourse on financialisation.

First, and perhaps most obviously, the domination of the economy by finance, with its dynamic of self-fulfilling promise and speculative logic, explains the *particularly* crucial role of promise, imaginaries, future visions, hype, performativity and of the cultural sphere more generally on capital accumulation and modes of regularisation at present (Brown and Michael 2003, Brown *et al.* 2000). The KBBE qua discourse thereby takes on a particular significance, for it is not just idle stargazing but an absolutely central element of the project to construct new possibilities for profit.

Second, just as the KBBE promises that knowledge and 'nature' can be the source of economic value, so too finance works on the premise that money invested in money can create profit and indeed wealth. Under conditions of financialisation, with enormous sums accruing to those working in finance, to deny this seems to fly in the very face of manifest fact. Yet, we have already seen that only productive labour can create surplus value. Under conditions of financialised neoliberalism in particular, however, finance has the political power to rearrange the *distribution* of value across the economy dramatically in its favour, even if its chosen investments do not themselves contribute much, or indeed any, new surplus value to the economy as a whole (Birch and Tyfield forthcoming). The impact on the credence given to, and hence social power of, KBBE discourse of the widespread observation and acceptance of money begetting money, however, must also not be under-estimated – especially in the context of the heightened significance to profit of social belief under conditions of financialisation, as discussed in point 1. For such common socio-political acceptance neutralises to a significant extent dissent at the inequitable distribution of wealth and growth in economic inequality that financialisation necessarily tends to create, thereby smoothing continuing capital accumulation.

Third, as the creation of many new knowledge and biotech industries is essentially explorative and, indeed, often speculative, financialisation is a crucial condition for the investment these need to be viable businesses, even in the short-to medium-term. The massive volumes of finance in these circumstances that is in search of return creates a market for the widespread investment in relatively speculative and innovative businesses (and particularly in technology, for which problems of private appropriation of profit are mitigated or at least diminished) with the promise of super-profits. The continuation of such investment, however, itself depends upon strategic calculation that accepts the KBBE vision, or at least accepts its widespread social acceptance.[9] The desperate and widespread search for maximum financial return also, in turn, affords finance the capacity to effect the violent changes of 'creative destruction' (Schumpeter 1976) to the real structure of the economy that are needed for any such vision to come to fruition.

Fourth, there has been much commentary about the KBBE, and especially biotech, which reveals that there has been little delivered in comparison to its grand promises and the massive investment with which it has been awarded (Arundel 2000, *The Economist* 2010c, Nightingale 2004). Yet, even if investor confidence in biotech were to collapse tomorrow, it has been continuing for nearly 30 years now (since the late 1970s). It is clear, therefore, that another feature of the KBBE discourse is its ability (to date) to ride the breaking waves of biotech hype and continually to dissociate itself from the aggregated evidence of individual business failures. This perpetual postponement of the reckoning of investment in the biotech sector in general, however, is also crucially dependent on financialisation and its political and cultural dominance. For it is in this context that it becomes intelligible that the promise of great riches will always continue to trump all the evidence to the contrary at the material level of the productive economy.

Finally, the current praxis of the KBBE involves not only the commoditisation of knowledge and 'nature' (as we shall see) but also their transformation into 'fictive capital' (Jessop 2007), in which the commodity's supposed future returns from its revenue stream are securitised, i.e. made into financial instruments that may themselves be bought and sold. This further complicates the relationship between KBBE and financialisation, because such securitisation, for instance via trade in intellectual property rights, is a major source of investment in the KBBE. Yet, it is clear that both the promise of substantial future returns (and hence the price of the securitised product) and the reasonable expectation that such securitisation will be possible *in the first place* (and hence there being a market for these financial products, hence investment being available via this route) are both entirely dependent on the KBBE imaginary (Orsi and Coriat 2006).

We have seen, therefore, that the KBBE qua discourse cannot simply be dismissed as so much empty nonsense, a mere 'buzzword' (Godin 2006). For while it is undoubtedly false in its central claim and at best hugely overstated as a possible vision of a future economy, it also plays the crucial role of engendering, cohering and mobilising sufficient political support for the ongoing project of finance capital (to attempt) to *construct* a profitable KBBE and to profit spectacularly in the process. However, it is also the case that, on the basis of this analysis, the continuation of the KBBE imaginary is hugely dependent upon the preservation of a financialised neoliberal (global) economy. The implications for the KBBE of the current upheaval in the financial sector, especially in the dominant nation-state economies centred on the US, and hence the global economy more generally, are thus potentially extremely serious and will warrant continued research. Indeed, recent reports have shown that the present crisis has precipitated a singular downturn in venture capital (VC), including in the US, upon which biotech funding is singularly dependent (e.g. *The Economist* 2010c: 7), concluding that, 'listed biotech companies, which make up 10 per cent of America's total stockmarket listings, are heavily dependent on external finance and their growth is likely to suffer far more from a withdrawal of credit than that of the overall economy'.

3.4 The real KBBE – the case of agriculture

With these preliminaries noted, we can now turn to a more detailed examination of the KBBE as it actually exists and is coming to be. Agriculture is a key area of the KBBE, alongside fisheries, forestry and biotechnology, according to official EU policy documents (see Birch *et al.* 2010, European Commission 2010). Examination of the trajectories of socio-technical and political economic change in this sector, therefore, provides an excellent illustration of the concrete impact of KBBE discourse and the political economic reality that is actually being constructed, including the roles, actual and envisaged, of (agricultural) science. As we shall see, the divergence between this reality and that envisaged by KBBE discourse could hardly be starker.

Agriculture has undergone seismic changes in the past 30 years, with the emergence of a global agri-food system (e.g. Bonanno *et al.* 1994, Friedmann 2005, Magdoff *et al.* 2000, Weis 2007). This 'corporate food regime' (McMichael 2009, van der Ploeg 2010) is characterised by:

- monopolistic/oligopolistic food supply chains dominated by TNCs (in inputs, food trade, processing and retail);
- an industrialised, high (fossil fuel-based) input, monoculture model of agriculture, especially in OECD countries;
- global trade prices and markets, dominated by exports from these OECD countries;[10] and
- commensurate neglect of localised agricultural resilience and distribution of consumption.

These fundamental characteristics have together conditioned the emergence of a longer-term agrarian crisis that, in turn, culminated in a full-blown food crisis in the food price spikes of 2007–08, precipitating food riots in more than 25 countries across the world (Bush 2010). Such social unrest has since died down for the time being, but there remain gross inequalities in the global distribution of food, with over 1 billion under-nourished on the one hand, and another 1 billion clinically obese on the other (FAO 2010, Lang *et al.* 2009, Tansey and Worsley 1995).

Moreover, further, and more serious, food crises are likely in future. For several trends (themselves with varying degrees of connection to neoliberal financialised globalisation and the KBBE) regarding agriculture and food security are coming together in the creation of what the UK's Chief Scientist has called a 'perfect storm' (Beddington 2009). These trends include: demographic growth outstripping growth of food production and yield improvements; increasing competition between food and biofuels for staple crops, particularly grains; increasing and chang- ing patterns of consumption, especially increased eating of meat, in massively populous developing countries, notably China and India; and a series of growing environmental challenges regarding the (unpredictable) effects of climate change, scarcity and pollution of irrigation water and shortages of critical resources (particularly phosphorus).

It is clear, therefore, that agriculture, far from being a strong and robust economic sector, is facing a gamut of extraordinary challenges. This has prompted significant recent policy interest in agriculture, with 'food security' rising precipi- tously up political agendas, including in the global North after several decades of relative neglect (Chatham House 2009, DEFRA 2009, Royal Society 2009). Moreover, the latest versions of KBBE discourse incorporate these concerns and advocate a programme that will resolve them by simultaneously tackling the triple issues of increased agricultural yield, increased profits and economic competitiveness and improved environmental sustainability.

This programme involves the reorganisation of agricultural economic activity, but how exactly is it being transformed? The actual KBBE in agriculture that is emerging to date is, in fact, marked by an archetypal neoliberalisation via the (attempted) four-dimensional real subsumption by capital of food production, agriculture, agricultural (scientific) knowledge and (agricultural socio-)nature. In each case, however, these four developments generate significant problems that, far from successfully allaying the underlying problems, seem to be directly contribut- ing to, producing and reinforcing agrarian crisis. And, just as we saw how complete commodification or valorisation of socio-nature and knowledge is impossible, so too the 'overflowing' of these subjugated realities in each case is itself condition- ing the emergence of novel problems that compound the instability of the global agri-food system, adding further conditions to already massively over-determined future food crises. Finally, through its growing command over agricultural practice, policy, research and innovation (R&I), each of the four developments also serves to marginalise (further) alternative visions for a 'KBBE' that may offer more effective ways to move towards more attractive futures. In short, while contemporary KBBE discourse marks a break with previous neoliberal discourses of the 'post-industrial' economy, there is a fundamental continuity in the changes to global agriculture. As such, in studying the current trajectory we may also understand how KBBE discourse, as the most recent incarnation of policy for a neoliberal, financialised globalisation has conditioned the agrarian crisis in the first place.

Our starting point in analysis of the neoliberal assault on agricultural is the prima facie resistance of farming, via seed (or livestock), to capitalist logic. At the heart of agriculture is the natural cycle of seed → crop → seed etc. Yet, through this process, in which the production of the crucial means of production is naturally and spontaneously reproductive, the seed offers the farmer a seemingly permanent means to evade the private appropriation and control of their economic activity by someone else, such as a capitalist. Hence, while agriculture, particularly agricultural land enclosures, has arguably been the first economic sector to be valorised in the historical processes of expansion of the capitalist economy, whether in a particular country (as per 'land reform') or globally (as per Britain's early modern Acts of Enclosure and the subsequent eighteenth-century 'agricultural revolution') (Wood 2000), it has also remained a sector that has been particularly resistant to wholesale capitalist reorganisation (e.g. Kloppenburg 1988/2004, Mann and Dickinson 1978, Mascarenhas and Busch 2006, Shiva 2001).

As a major sector of economic activity, with fairly inelastic (if not guaranteed) demand, however, it has always been a particularly attractive sector for capital from which to attempt to profit. This has led to the emergence of a number of developments that attempt to circumnavigate the obstacle of the seed's natural reproduction and successfully subsume agriculture to the logic of capitalist accumulation. All four have been particularly marked during the last 30 years of neoliberal and financialised globalisation, though as we progress from 1 to 4, we shall see that the developments become more specifically contemporary and KBBE (or equivalent) discourse also becomes more influential.

1. Real subsumption of the agri-food supply chain

First, the earliest and socio-technically most straightforward development has been simply to decompose the overall process from seed to plate into as many steps as possible and to shift as many of these away from the non-capitalist farm as possible. In many, if not most, traditional societies, this entire 'supply chain' will have generally been within a single household or rather a single community working land held in common. Production and consumption are thus first separated by the dispossession of such common land (which may itself create a landless mass that provides the necessary workers for a capitalist industrial economy), a process that is far advanced (but still not yet finished) in the rich global North. This also creates the great 'metabolic rift' (per Marx, as quoted in McMichael 2009) of the modern era, between an agriculturally productive countryside and food consuming towns, that breaks traditional local and small-scale cycles of waste reuse and engenders instead massive concentrations of waste as pollution.

However, further subdivision of agriculture is a much more recent development, and one that has been particularly marked in the past 30 years. This involves a dual process of simplifying and standardising farm processes and their subsequent integration into the broader industrial economy, both upstream (regarding farm inputs) and downstream (regarding farm products), hence maximising capitalist control of agriculture, even while growing itself remains resistant to being organised in that way (Bernstein 2001, Kloppenburg 1988/2004). For both of these processes facilitate the transformation of multiple steps (including possibly new ones) in the agri-food supply chain into sites where capitalist relations of production can be (more easily) imposed given various specific conditions. Moreover, as regards an economics of science, the innovation and transformation of these (now) off-farm production processes often incorporate considerable scientific research; consider, for instance, the army of food scientists at major food processors, such as Unilever or Kraft, working on every step of the agri-food supply chain.

The result of this process has been the massive industrialisation of agriculture. Upstream, agriculture has become increasingly dependent upon multiple inputs, particularly mineral (i.e. fossil fuel-based) fertiliser and biocide treatments that can themselves be (and are) manufactured in industrialised and fully capitalist production processes. The mechanisation of farming has also inserted the need for industrially produced and serviced machinery in order for a farm to remain competitive and economically viable. Similarly, downstream, processing of raw farm produce has been shifted off the farm to industrial-scale factories – the growth in such labour offsetting to some extent, in aggregate and quantity at least, the huge reduction in labour needed on the farm itself as it has been mechanised – and farm ownership has been more concentrated in search of economies of scale.

In short, therefore, these developments are all examples of the attempted real subsumption by capital of the agri-food supply chain. This has taken two forms. First, as much of the agri-food production process as possible is transformed from *unproductive* labour from the perspective of capital (in that the farmer is not a wage labourer employed by a capitalist) into PLK. Second, this shift of the agricultural labour process off-farm into industrial settings in turn affords the familiar methods of disciplining a workforce to maximise relative surplus value from labour through reduction of the socially necessary labour time and turnover time.

Each of these developments is in ample evidence in the current agri-food sector. Thus, as regards the increasing intensity of labour exploitation, post-farm food processing factories, such as abattoirs, are among the most intensely time pressured and, consequently, dangerous work environments in developed economies (Sun and Escobar 1999). As Weis (2007: 84) notes, 'an average US chicken slaughter and processing plant will kill and refine into packaged meat 150,000 chickens in a single shift – or 300,000 chickens every working day.'

Similarly, as the agri-food chain has shifted off-farm and its profitability has, in turn, increased (global) competition, it has become increasingly concentrated in the hands of a few giant TNCs (mostly based in the US and EU). As Weis again notes (ibid.: 73):

> In 2004, the top ten TNCs controlled 84% of the US$35 billion global agro-chemical market (led by Bayer, Syngenta, BASF, Dow, Monsanto and DuPont); the top ten TNCs controlled roughly half of the global US$21 [billion, sic] global commercial seed market (led by Monsanto, DuPont/

Pioneer Hi-Bred and Syngenta); and the top ten TNCs controlled 55% of the US$20 billion global animal pharmaceutical market, the great part of which is used in animal agriculture.

In short, the real subsumption by capital of much of the agri-food production process has been spectacularly successful, at least on its own terms. It has also, however, generated numerous significant problems, including for further capitalist accumulation itself.

First, the industrialised and high-input model of agriculture these changes have entrenched have trapped agriculture in a 'technological treadmill' (Cochrane 1993), the familiar predicament of capitalist industries. For instance, farmers must perpetually upgrade farm machinery in order to remain competitive. Perhaps the most serious example, however, is the 'pesticide treadmill'. Modern industrial agriculture is not only dependent upon purchased chemicals produced industrially off-farm but must continually increase its use of pesticides, both to keep up with competition by maximising crop yields and to respond to emerging pesticide resistances, against steady crop losses of 25–40 per cent on average (Birch 2010). Not only are huge sums thus spent on treatments of ever-diminishing efficacy, but the increasing intensity of these chemicals also has significant costs (both financial and ecological) in terms of pollution.

Moreover, the continual development and obsolescence of pesticides is now confronting limits that threaten this entire model. Approximately 60 per cent of known pesticides are now useless and a co-evolutionary tipping point is now being reached in the pesticides 'arms race', where pesticide resistance is emerging faster than new pesticides can be developed (i.e. approximately 10 years) (Birch 2010). This is particularly significant for the EU, which, despite the vaunted sustainability of KBBE discourse, is the largest producer, user and exporter of pesticides.

Second, as upstream and downstream processes have become industrialised and concentrated, farming itself has been increasingly squeezed from both directions by the intense cost pressures of globalised competition that these more economically powerful agents can simply pass on to the farmer. For instance, in their ever-greater attempts to control the agri-food supply chain and to divest risk and externalise costs, Wal-Mart have recently developed a system by which the farmers from which it buys fresh fruit and vegetables will only be paid once their produce is scanned at a till in one of its supermarkets. All the risk associated with transport and storage of such perishable produce is thus to be borne exclusively by the farmer.

The continually intensifying pressure of developments such as this thus leads to ever-decreasing margins, a financial 'race to the bottom' (Marsden 2003), which is increasingly making farming economically unviable without various state handouts. This, in turn, has various effects that undermine the future of agriculture, including: a rapidly aging farming workforce as the relentless stresses upon agriculture repel younger generations from staying on the farm; and, conversely, increasing dependence of on-farm production on a 'super-exploited' workforce of temporary (and often immigrant) labour that faces low wages, poor and often hazardous working conditions and limited benefits or job security (Majka and Majka 2000: 163; Weis 2007: 84).

Furthermore, with margins so tight or negative and competition so intense, the only way to pay for the necessary machinery upgrades is to assume ever-higher levels of debt. This, in fact, is itself a further way to subsume the farm within the cycle of capital accumulation, as servicing the debt effectively renders the farmer the bank's employee. But it has become apparent in the aftermath of the 2008 financial crash that the indebtedness of farmers in many countries with industrialised agriculture is now at unsupportable levels. For instance, a rise of 33 per cent in debt levels between 2002 and 2007 left Dutch farms with total debts of €38.8 billion, equalling 1.7 times of the total gross value of agriculture produce, 6 times more than total value-added and 12 times more than total farm income that year (van der Ploeg 2010: 105, quoting Berkhout and van Bruchem 2008). Nor are Dutch farms exceptional, with 'considerable segments of the agricultural sectors in the US, Latin America, South Africa, some parts of Asia, Eastern Europe and the rest of the European Union . . . showing (albeit sometimes for different reasons) similar or even higher levels' of debt (ibid.). There are thus considerable risks of mass default, with the sudden closure of multiple farms and/or a flood of livestock sales and slaughter, producing a commensurate shock to food production that would ripple through the global agri-food system.

These debts also heighten the dependency of farms on (global) capital markets (van der Ploeg 2010: 100), which in turn increases the sensitivity of agri-food production to the vagaries of what are, under conditions of financialised globalisation, increasingly volatile global markets and economic trends (e.g. Wade 2006). In sum, therefore, the real subsumption of the agri-food system has fundamentally destabilised food production in the geographical centres of globalised, industrial agriculture.

2. Commodification of agriculture

The next set of issues we will consider are those regarding the increasing penetration of capital and intensified labour pressures into processes *on* the farm. Two developments stand out, in both of which (scientific) knowledge plays a key role (hence being particularly important for an economics of science). One of these, the development of genetically modified (GM) crops (and animals), is now the subject of a burgeoning literature, both academic and popular, but the second, supply-chain management, has received much less comment.

Agriculture actually produces *two* products: the commodity produce that it sells for consumption and the seed, the means of further production. It is regarding the *latter*, however, that agriculture exceeds subsumption within cycles of capital accumulation, as discussed above. The classic process of formal and real subsumption of industry, as described by Marx (1992, 1999) involves, respectively, the dispossession of workers of the means of production and their aggregation into factories, on the one hand, and the subsequent and seemingly limitless process of incremental intensification of labour productivity (hence increasing relative surplus value), on the other. Both of these processes are seemingly difficult (if no longer 'objectively' impossible, as Mann and Dickinson could plausibly argue in 1978) in the case of agriculture:

α   Farms (and farmers) are immobile and so cannot be easily aggregated into panopticon factories where they may be subject to capitalist discipline. Moreover, the natural reproductivity of seed and its inseparable production from the production of commodity produce makes appropriation of the means of production extremely difficult to enforce. This therefore interferes with the cycle of capitalist valorisation.

β   Second, natural cycles of growth make it difficult to intensify the labour process, for production of both agricultural produce and seed alike. This thus undermines the (medium-term) profitability of seed production, in particular, and hence also the motivation to appropriate this process in the first place, even if it were easy.

Taken *together*, therefore, these points undermine the formal subsumption of agriculture while, given formal subsumption, point β) would be the focus of attempts at its real subsumption.

In fact, however, we have already seen various ways in which, at least as regards the on-farm production of *commodity produce* (i.e. not seed, though this may be treated as inseparable for the time being, hence the complications of agriculture for capitalist reorganisation), agriculture *can* be effectively valorised. Thus, the dispossession of farmers, either through a shift to tenancy farming or bank debt, while not strictly making the farmer a waged labourer, does transform their economic situation into one of earning a performance-based commission on surplus produce over a given target (needed to meet their rent and/or interest payment). Similarly, we have seen how work on the farm is already arranged as waged labour producing commodities for sale in search of profit; and, indeed, constitutes a 'super-exploited' workforce. The industrialisation of agriculture, discussed above, also shows that agricultural production can be, and has been, dramatically intensified without the need to tackle directly points α and β above.

Nevertheless, the control and incorporation of agriculture itself into cycles of capital accumulation is still incomplete and the two issues discussed in this section are both significant steps in this direction. In fact, neither development is strictly the subsumption of agriculture under capital, in that they do not involve the conversion of a production process into waged labour (as do the other three developments). They do, however, both concern increased *control* of agriculture within cycles of capital that, in turn, facilitate the (future) formal subsumption of agriculture. Both also, therefore, provide new ways to valorise and profit from agriculture itself and give rise to a difference between formal (e.g. legal) and real (e.g. technological) control by capitalist firms, with a progressive shift from the former to the latter, is clear.

The first issue is the emergence of supply-chain management (SCM). As rich world consumers have become increasingly demanding and varied in their tastes, and food processing has driven down the margins in basic foodstuffs, competition in food markets has increasingly migrated from price reduction to qualitative differentiation and branding (Busch 2007). This has created the demand from major firms in the agri-food supply chain for the development of quality assurance mechanisms and standards. This, in turn, has had dramatic effects on the farm, for as these techniques have become more sophisticated and detailed they have also increased the control of downstream firms over on-farm processes via the 'objective' quantification of qualitative properties of agricultural produce. As such, these novel knowledge technologies, which are increasingly techno-scientific, afford the *construction of value* by effectively expanding and creating new markets.

Recall that 'value', in the strict sense of a capitalist economy, is not 'real' (in the sense of self-subsistent or existing independently) but is dependent on social (cultural, political, technological) (per)formation, in particular of the means to 'objectify' comparison of commodities (products, services) in market competition and the production processes underpinning them. Value, thus, must – in all cases, and not just for agriculture – be *constructed*, depending on the (techno-scientific) development of such means, though this process may be more or less straightforward. Hence, the Industrial Revolution itself depended upon the development of clocks to measure 'factory time' or railway timetables (e.g. Glennie and Thrift 1996, Thompson 1967, Urry 1995), as well as techniques for measuring factory output, efficiency and productivity, and accounting techniques (e.g. Porter 1995). Today, however, the need for such knowledge technologies for valorisation is starkly apparent for agriculture.

Moreover, SCM thereby creates new opportunities for profit. The value of a commodity reflects the socially necessary labour time of its production, in competition with other firms producing the same commodity. But what is the 'same' here? SCM affords ways to 'objectify' different 'classes' of a previously homogeneous commodity. This then allows the winners of this process of market segmentation (who, of course, drive it forward) to gain through the super-profits associated with their higher 'quality' product. As with the technological treadmill, however, such gains are only ever short-lived due to competition. In the medium-term, therefore, SCM intensifies the already intense competitive pressures it was designed to overcome, as the parameters of competition become more and more sedimented. This, in turn, increases pressure on the farmer (and the agricultural worker in particular) to increase productivity (both quantity and quality) in order (for wages) to be financially viable within these diminishing margins.

Moreover, this increased control over agriculture can be 'formal' or, increasingly, 'real'. Increased formal control would include such ideational, cultural, policy and/or legal changes as standards and branding. And, indeed, standards for agricultural produce, usually established by private foundations or corporate alliances rather than public agencies, have proliferated in recent years to the point where such private regulation dominates the agri-food system. But real control is also increasingly afforded by the development of new techno-scientific knowledge, such as science-based metrics, that render the farmer simply unable to sell their produce (profitably) where their production process and produce does not pass seemingly unarguable machine-enabled tests (Busch and Bain 2004: 15).

While SCM concerns the increased control of agricultural production per se, our second issue, GM crops, relates to the control of the agricultural means of production. The importance of this development has been perhaps best captured by Kloppenburg (1988/2004: 201), when he writes of the seed as '*the nexus of control over the determination and shape of the entire crop production process*' (original emphasis) (see also Mascarenhas and Busch 2006, Shiva 2001, Weis 2007). For, as we have seen, insofar as GM facilitates the private appropriation (and so accumulation by dispossession (Harvey 2003) or primitive accumulation (Marx 1999)) of the naturally reproductive means of production of agriculture, it directly tackles one of the two persistent obstacles for agriculture's valorisation (point α above). Since effective formal (let alone real) subsumption of agriculture also demands the potential intensification of production of the means of production (β), however, this is merely a necessary but insufficient step towards such subsumption. Sure enough, however, attempts to tackle the second point are also underway, as we shall see regarding the fourth dimension of the agricultural KBBE below.

New techno-scientific developments, particularly the biotechnological capabilities for genetic modification, are the key to this process; arguably of even greater importance than SCM. For it is the emergence of these new sciences that has enabled the effective appropriation of the seed, and in two ways. First, formal control, via legal ownership, becomes possible and easier with GM crops because the genetic modification allows a firm to claim intellectual property rights (notably patents) over the new variety and so prohibit use of its seed without payment of a licence fee. The effect of such exclusion has been significantly enhanced by a number of contemporaneous (but not coincidental – see Chapter 4) and quint-essentially neoliberal developments in intellectual property rights (IPR) law and practice, both nationally (especially in the US) and internationally (via the World Trade Organisation (WTO)), that has strengthened, extended and increased the penalties for breach of such patent rights.

Second, real control of the means of production, hence not relying just upon the leaky and potentially expensive mechanism of legal exclusion alone, has also been made possible by these new biotechnologies. This can take place either by developing crops that must be used, if they are to maximise yield in the context of the extreme competitive pressures, because they are resistant to or compatible with a package of jointly sold inputs (e.g. biocides); or by the development of crop lines that are effectively infertile and so have to be purchased anew for each growing season, so-called 'terminator technology'. In both cases, therefore, GM technology affords the development of new forms of PLK through the construction of new markets (or rather conversion of existing labour processes) via new material possibilities.

Both of these developments, however, are resulting in the emergence of new problems for agriculture, so that they are, at most, merely *attempts* at establishing real control. First, regarding SCM, in search of (temporary, degenerating) super-profits, private standards in the agri-food supply chain have proliferated. Yet, these standards are often incommensurable, producing a tangle of regulation and a consequent fragmentation of agri-food supply chains (Busch 2010). This is thus imposing significant management costs, which the more powerful players seek to offload but which even they cannot completely escape, hence undercutting the (economic) rationale for the development of such standards in the first place.

In the case of GM, on the other hand, while such technology has definitively shown that no 'natural' and 'objective' limits to agricultural production can be taken for granted, socio-nature nevertheless remains recalcitrant and capable of surprise, necessarily exceeding contemporaneous scientific understanding at any given time (an actant in its own right, in the language of actor-network theory (ANT)) (Castree 2002, Latour 2004, Smith 2007). Moreover, this is especially true when set against the one-eyed focus of capital accumulation on exchange value. For this thereby goes beyond, and exacerbates, the inescapable predicament of the limits of knowledge of agricultural nature by forcing such knowledge into the Procrustean bed of the particular needs and expectations of maximising production of surplus value, hence resulting in 'massive quantitative and qualitative dislocation' (Castree 2002: 138–139). For instance, to maximise profits, agricultural socio-nature is: extracted from its indissociable ecosystemic context; subjected to the alien and short-term temporal horizons of industrial production; exposed to a relentless dynamism and change against the relative stability of life cycles; and reduced to the homogeneous quantitative measure of SNLT (ibid.). Nor is there any reason to expect that the development of science-based metrics for SCM will evade similar problems of (increasing) complexity and unmeasurable or 'overflowing' realities.

In these circumstances, it may be little surprise (if inadequate reassurance) that the sheer difficulty of GM technology has also proven to be a major problem. GM for most traits has turned out to be extremely complex, hence with limited success, prompting many to downgrade the 'biotech revolution' to mere 'myth' (Arundel 2000, Nightingale 2004). For instance: getting a good 'event' (i.e. a useful trait) is slow and expensive; the technology is best for simple traits but it has proven hard to find these; subsequent stewardship of the crop as it is developed is critical but costly; and maintaining stocks for multiple events is a significant burden on breeding programmes (Lawrence 2010). Indeed, as a senior executive from a major agri-biotech TNC has baldly put it, the main significance of GM is not its direct successes but that it changed the commercial model, making breeding profitable (ibid.).

All these problems of limited scientific understanding and control, however, have obviously inflamed GM as a hugely controversial political issue. At its height in the late 1990s and early 2000s, this debate focused on issues of the importance or not of a 'precautionary principle' and the 'substantial equivalence' or not of GM crops to existing (and already regulated) lines (e.g. Bauer and Gaskell 2002, Jasanoff 2005, Levidow *et al.* 2007, Stirling 2001). In both cases, we should also note, science had a key role to play, adding a further angle to an economics of science. A crucial (and arguably more productive and reasonable) aspect of this debate, however, was not criticism of 'GM' per se, treated as a single, monolithic technology, but the associated corporate control of the agri-food chain. To date, 'defenders' of GM continue to argue that it is a technology that is absolutely essential to increase global food security and feed the poor in the global South, where environmental conditions are becoming increasingly hostile to conventional crops. Yet, the supposed drought- or saline-resistant GM crops are conspicuously absent. Instead, the vast majority of GM crops on the market, and investment in research for future crops, either offer compatibility with other (biocide) inputs sold by the same TNC or are terminator technologies (Weis 2007: 72); i.e. precisely the two mechanisms to attain real control over agricultural production.[11]

Moreover, since GM 'made breeding profitable' the concentration of the agri-food sector in the hands of a few TNCs has increased significantly (see above). With these firms then particularly dependent on global trade, a crucial aspect of the GM controversy has thus been the transatlantic quarrel regarding their safety (Winickoff et al. 2005), with a (reluctant) European Commission being compelled by popular dissent to stall GM agriculture in Europe while the US, dominated by the agri-science TNC lobby, threatened trade sanctions in retaliation.

Finally, both SCM and GM technologies also have significant implications that threaten to exacerbate the existing conditions of agrarian crisis. In particular, by dominating the process of agricultural production and concentrating financial muscle, both processes exert considerable power conditioning the trajectories of the development of agricultural knowledge and innovation. This is not just as regards the direct effect of development of the knowledge technologies needed for each process, but also via indirect effects. Hence, through their domination of the agri-food system, private standards now set the research agenda regarding the breeding of particular traits that will meet market demand. Similarly, the science-based metrics being developed for SCM are becoming increasingly crucial preconditions of invest-ment in innovation in order to enable the differentiation of innovative seeds/produce that are proprietary from the rest (Busch 2010). Both of these develop-ments, therefore, are conditioning a transformation of agricultural innovation, to which we turn next.

3. Real subsumption of agricultural innovation

The goal, and often effect, of the changes we have discussed so far is to create and deepen the dependency of the farmer on the agri-food TNCs, with various versions of the technological treadmill as a major mechanism. The separation of farmer and the production of farming inputs, however, also affects the separation of farmer and *development* or *innovation* of these means of production. Yet, the treadmill accelerates socio-technical change and so intensifies the dependency of the farmer on innovation provided for sale off-farm yet further. Moreover, this innovation is increasingly of a *particular model*, namely privatised, R&D/

biotech-intensive (and so expensive) development by TNCs of hi-tech commodities (such as GM crops). Such innovation products are then patented or otherwise protected by IPRs, hence promising super-profit monopoly rents in return. This, then, is the supposed route to ensure the 'competitiveness' of the EU in KBBE discourse.

It is important to consider the conditions that are driving this new model of agricultural innovation and the imperative of super-profits of innovation. One crucial aspect is the increasing globalisation of agriculture, as the dominant TNCs seek to expand markets in search of greater returns; a process that then conditions and is conditioned by the further global concentration of these industries, through mergers and acquisitions. The result is that agriculture (and particularly big agri-business of the 'grain-livestock complex') is among the most sensitive industries in the US to global trade, 'twice as dependent . . . as the US economy as a whole' (Weis 2007: 69). In these conditions of global competition among enormous TNCs, however, pressures on profitability are even more intense, placing a premium on proprietary innovation. This, in turn, unleashes the self-propelling dynamic through which global competition drives hugely expensive R&D programmes (which must be funded, hence bringing in high finance) in search of patent monopoly rents, which then must be sold on a global scale in order to recoup investment, so driving further global integration and liberalisation of agriculture and even more fierce global competition.

Moreover, as science-intensive innovation, this model also draws agricultural innovation into a 'global KBE treadmill' (see Figure 3.1). While (techno-scientific) knowledge-intensive innovation cannot be presumed to be universal, it is not an unreasonable strategic business assumption that it may be potentially global in source (e.g. new biotech products could, in principle, be developed anywhere that there is a strong research system) and consumption/application (e.g. they could be sold and used anywhere without significant local adaptation). Accordingly, science-intensive innovations are both singularly attractive in their promise of patentable innovation with potentially global markets *and* especially amenable to increased global competition, hence the particular emergence of global innovation networks in such knowledge-intensive industries (Ernst 2008).

Furthermore, these seismic changes in the social locus of agricultural innovation into corporate R&D laboratories are occurring alongside and in co-production with the privatisation of (agricultural) science more broadly; our starting point in our inquiry into the economics of science. While conventional 'economics of science' analysis thus focuses exclusively on this latter phenomenon, we see clearly, therefore, how it is just one part of much broader and systemic transformations in the political economy of R&I; and, indeed, unintelligible in the absence of this contextualisation.

Three phenomena illustrate the ongoing privatisation of agricultural science. First, the neglect and decline of public agricultural research across the developed global North is apparent. This includes the downgrading or abolition of public research institutes (PRIs) and public extension services. Indeed, 'the public sector

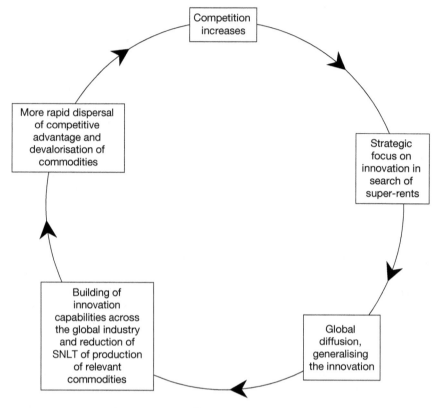

FIGURE 3.1 The global KBE treadmill[12]

no longer has adequate financing and human resources to educate and train plant breeders', but is essentially now tasked with 'transferring traditional knowledge and training to the private sector' (Mascarenhas and Busch 2006: 6, quoting Frey 1996). Private sector funding of agricultural research, conversely, 'tripled in real terms between 1960 and 1992 (Fuglie *et al.* 1996). By 1994 the total estimated expenditure for plant breeding in the US was $551 million, of which 61 per cent or $338 million was from the private sector (Frey 1996)' (ibid.). Of course, this process has involved the massive conversion of knowledge labour that was previously unproductive for capital into PLK. Finally, there has been a massive parallel rise in agri-biotech patents in this period, annual US patents granted for plants rising from 318 in 1990 to 1,240 in 2008, or an average of 407 per annum in the 1990s to 945 in the 2000s (data from USPTO). And, as public funds have diminished, universities and research institutions have changed their research agendas to focus on areas that may win funding from the private sector. Two particular egregious examples of this in the UK have been the closing of world-leading research institutes in forestry (at Oxford) and horticulture (at Warwick).

In short, therefore, these trends are simply the multi-pronged attempt to effect the real subsumption of agricultural (scientific) *knowledge(s)*, with the classical three stages of this process apparent: first, the appropriation or accumulation by dispossession of traditional agricultural knowledge using IPRs (or 'biopiracy'); second, formal subsumption by the separation of farmer and innovation, with the latter (given the commercialisation of science) transformed into PLK; and finally, real subsumption through the development of new techno-sciences (and associated innovation) that farmers cannot possibly do for themselves, and that cement their total dependence on external capitalist innovation to keep the farm economically viable.

But, again, these are merely an 'attempt' at such real subsumption, and new problems are emerging. First, a series of problems arise as regards the paradoxical and unintended effects of attempts to increase or establish control of reality on the basis of scientific certainty. These include a disconcerting creation of further sources of uncertainty and risk with each seeming advance of epistemic control (Beck 1992). While corporate R&D thus seeks to develop scientific knowledge that will assure ever-greater control of its production process, this is persistently undermined by the parallel proliferation of uncertainties. Furthermore, while science is repeatedly developed/funded in the hope that it will provide objective certainty, in fact, the more important science becomes to socio-economic life, the more contested it becomes. This is not just a matter of contestation of direct techno-knowledge-based economic interventions (as per GM in EU), and especially as developed by corporate R&D, and the 'regulatory science' (e.g. Irwin *et al.* 1997) providing evidence for policy as to their effects (or absence thereof). But also the increasing importance of such regulatory science as a means for the construction of value and the regularisation of capital accumulation on the basis of 'objective' scientific measurements and rules.

Perhaps the most striking examples of these processes are GM crops, where the attempted closure of political debate about their safety through mobilisation of supposedly definitive scientific evidence led to the very opposite outcome, not only failing to depoliticise the controversy but backfiring, with the politicisation of the science (Levidow *et al.* 2007).[13] While some scientific knowledge is indeed sufficiently sedimented and assured that this would not always be the case, it is also unlikely that such widely accepted 'fact' would be ever be at issue in such a political controversy.[14] Rather, the political use of science (including by large corporations) to end dissent is likely to involve science that is itself in a state of considerable flux and disagreement among disciplinary experts themselves; this is, after all, the *norm* for science at the 'cutting edge', the very source of its epistemic productivity. Political (or commercial) judgement based upon such science will thus simply expose the sheer *uncertainty* and disagreement over the 'facts', while simultaneously introducing the dynamics of heated political debate. Given the evident growing uncertainties and scientific complexities of issues associated with agriculture following such hi-tech innovation, these problems are likely to characterise agricultural innovation for some time. In sum, therefore, as agriculture is integrated into the very heart of the 'knowledge-based economy' it is also, to that extent, increasingly exposed to its contradictions and aporia.

The second major set of problems includes those associated with the effects of the privatisation of research. First, we may note a number of the more general problems associated with a knowledge-based economy as they are relevant to agriculture. These are multiple and diverse. Jessop notes three intrinsic problems of capital accumulation that are especially significant for knowledge-based accumulation, namely:

> A dissociation between abstract flows in space and concrete valorisation in place; second, a growing short-termism in economic calculation vs. an increasing dependence of valorisation on extra-economic factors that take a long time to produce; and, third, the contradiction between the information economy and the information society as a specific expression of the fundamental contradiction between private control in the relations of production and socialisation of the forces of production.
>
> (Jessop 2000b: 68)

These trends are, indeed, in evidence regarding agricultural research, for instance: devaluation and neglect of longer term research while high risk and patentable research is overvalued; further dissociation of scientists and farmers, so that the former cannot know the needs of the latter; and a reduction of research spillovers and the cumulative productivity of knowledge through increased IPR claims (Busch 2010). There is a significant distortion and fragility (what Shiva (1993) usefully calls 'monocultures of the mind') in contemporary agricultural R&I. The real subsumption and monopolisation of agricultural knowledge, however, is also simultaneously undermining alternative, more robust and traditional forms, introducing a potentially catastrophic superficiality to global understanding of agriculture. As Mascarenhas and Busch (2006) have noted, it is the height of scientistic arrogance to assume that the traditional knowledge of managing seeds is a primitive and ham-fisted imitation of that embedded in GM crops. Rather, significant skill is needed in the choosing, storing, replanting and improving of seed, knowledge that has been accumulated and refined over millennia. As for any body of knowledge, however, this must be maintained, developed and transmitted if it is to remain a vital social capability, yet the real subsumption of agricultural knowledge is directly and deliberately subverting this process. Such is the sheer scale of the gamble involved in current attempts to shift agricultural knowledge production into corporate R&D labs.

4. The real subsumption of agricultural (socio-)nature

In point 2 (p. 75 ff.)we discussed how the formal, and then real, subsumption of agriculture is only possible when there is the possibility of intensifying the agricultural production process, for it is only in these circumstances that appropriation will not lead just to a short and one-off windfall but can underpin repeated rounds of super-profits as incremental improvements in productivity are innovated. The fourth and final development, therefore, is precisely this process. The emergence of a KBBE is thus marked not just by the valorisation and commodification of all stages of the agri-food process but also by the *intensified* capitalistic exploitation of natural resources in agriculture.

Moreover, while sections 1 to 3 are associated with the longer-term trajectories of neoliberalisation over the past three decades, this final process is the particular novelty evident in contemporary KBBE discourse. For this policy narrative envisions a three-fold path to sustainability via (i) biological = living/autopoietic = renewable and limitless resources, (ii) increased resource efficiency and (iii) unlimited innovation; i.e. with 'nature' (issues i and ii) and 'knowledge' (issue iii) as unlimited sources of not just a 'sustainable capitalism' but also 'sustainable capital', profiting from the autopoietic and hence supposedly limitless characteristics of knowledge and nature (Birch *et al.* 2010).

The slower emergence of this trend as against those discussed above may be understood simply as reflecting the extent of the technological challenge it involves. For the goal is to be able (continuously and recursively) to reduce the naturally necessary production time (NNPT) of agricultural production in order to increase productivity of the associated labour and hence ramp up the production of relative surplus; and such intervention evidently demands significant technoscientific capability. In this case, therefore, the neoliberalisation of (agricultural) nature goes beyond the construction of nature as privately appropriated 'resource' through accumulation by dispossession, as discussed above (and in Chapter 4) regarding IPRs. This latter process involves defining a particular socio-nature as 'property' according to a particular cultural-political conceptual scheme; a process at least as old as capitalism itself in the form of land enclosures and (colonial) seizure of mineral deposits. This is the 'formal subsumption of nature' (Smith 2007), for just as formal subsumption affords only the growth of absolute surplus value, so too this process is accompanied by the 'continual expansion of the conversion of extracted materials into objects of production' (Smith 2007: 13).

Conversely, what is now being attempted in agriculture is the 'real subsumption of nature', which aims to 'rework natural resources through new knowledge and technoscientific developments, which are themselves built upon the logics, strategies and expectations of capitalist accumulation through state-led (and –protected) market exchange' (Birch *et al.* 2010: 2901). This is only possible, however, in the context of the private appropriation not only of agricultural natures, *but also* of agricultural knowledge (point 3 above). Hence, the 'neoliberalization of nature is tied to neoliberalization of knowledge' generating a 'techno-knowledge fix' (ibid.).

This real subsumption of nature involves a two-sided shift (Smith 2007: 11). On the one hand, *capital* has always circulated *through nature*, but with its real subsumption in a 'KBBE', this circulation becomes intensified and strategically deliberate rather than just an 'incidental' and often frustrating side-effect. On the other hand, and vice versa, the circulation of *nature through capital* also becomes the strategic choice of capitalist business, and in two forms: as biotech commodities and/or as financialised products. Indeed, as Smith (2007) notes, the development of such biotech products has been accompanied by their almost immediate securitisation for financial markets (as per 'fictive capital' above), which in turn has been a significant source of investment, thereby illustrating again the tight connection between these processes of construction of an actual KBBE and the broader financialisation of the political economy, discussed above.

In fact, the process of intensifying natural productivity has been 'perfected' rather than invented anew with the advent of the new biotechnological capabilities. Hence Weis (2007: 19), dubbing the factory farm the 'apex of industrial agriculture' (2007: 59), notes that:

> From 1965 to 2005, the population of pigs on earth at a given time in a year has nearly doubled, reaching 961 million, and the population of poultry more than quadrupled, from 4.2 to 17.8 billion, but because the 'turnover time' of animals has been so shortened by industrial techniques the annual number of animals slaughtered has actually grown at a higher pace. Under industrial techniques, broiler chickens can be brought to slaughter weight in a few months, pigs in as little as six. In 1965, roughly 10 billion farm animals were slaughtered; by 2005, this had risen to more than 55 billion, led by a more than a sevenfold increase in the annual number of chickens slaughtered, from 6.6 to 48.1 billion.

Yet, there is a qualitative difference in the introduction of new biotech techniques, not least in that they enable the acceleration not just of the natural productivity of *agricultural commodities*, but also of *seed*, i.e. the agricultural means of production. Only when this 'upstream' production process is thus completely valorised is the real subsumption of agricultural nature finally possible. And this is a much greater technological challenge for crops than for animals, given the relative separation of reproductive and life cycles in the latter while crops only yield seed when they thereby also yield produce.

Once again, however, this attempted real subsumption is generating (and can be expected to generate) novel problems that illustrate intrinsic contradictions and limits of such development and/or exacerbate, rather than mitigate, the existing agrarian crisis. First, we may note that the dynamic of capital accumulation itself demands (as necessary presupposition of the valorisation process) that there are 'objective' limits (at least at the given time of realisation of value or sale) to the production process. The real subsumption of nature is necessarily a dynamic process that can only ever but must also continuously *redraw* the boundaries of socio-nature, and so can never settle on a 'sustainable' or stable end-state, 'completing' the capitalist colonisation of nature.

This point, while abstract, is nonetheless crucial when set against the supposed sustainability of the 'new economy' as discussed in contemporary EU KBBE discourse. The foregoing analysis shows the 'sustainability' of this KBBE discourse is (at least) doubly misleading in that (1) the implicit promise of long-term stability and limitless renewal held out by such discourse is entirely false, it offering instead (2) a vision of the 'sustainability' that entails ever more Promethean attempts to sequester socio-nature to the demands of a system whose logic is patently at odds, or rather intrinsically oblivious to, the innate, diverse and complex needs of bio-ecological systems. In these circumstances, the impossibility in principle of total colonisation of nature by capital may be scant consolation indeed, as mentioned above. It is likely, however, that more concrete problems will emerge to challenge the model before it gets much of a chance to put this to the test.

In particular, the recalcitrance of nature and the difficulty of engineering increased NNPT must be expected, just as is already in evidence regarding GM crops. Moreover, the particular attempt to intensify natural productivity may result in various forms of 'backlash' or nasty surprises, such as the emergence of diseases that already afflict livestock factory farming. Indeed, already, one of the major off-farm inputs in modern meat farming is pharmaceuticals, with US livestock now consuming a staggering 'eight times more antibiotics by volume than the human population', a rise in average intake by 50 per cent per farm animal and 300 per cent per poultry bird between 1985 and 2000 (Weis 2007: 72, quoting Mellon *et al.* 2001 and Nierenberg and Mastny 2005). This intense use of antibiotics, in turn, would seem to have obvious implications for the acceleration of co-evolutionary 'warfare' and hence both animal-borne diseases (e.g. avian or swine flu) and antibiotic resistance.

On the other hand, there is a striking lack of evidence that these innovations will lead to more efficient and recyclable use of resources. Rather, whether regarding irrigation water, fossil fuels, nutrients, soil, toxicity or energy (measured in terms of 'energy return on energy invested' (EROEI)) (e.g. Gregory 2010, van der Ploeg 2010) current trends in globalised industrial agriculture are all towards *increased* resource use and dependency and a commensurate worsening of environmental crises that are, in turn, aggravating the agrarian crisis. This is even, or especially, the case where innovations have improved resource use efficiency, for the effect of capitalist imperatives is to capitalise upon such improved productivity to increase overall production at the individual firm level, which in turn leads to increased use over the system as a whole; a feature not just of industrial agriculture but capitalism more generally.

Thus, while some have posited an 'environmental Kuznets curve', according to which environmental pollution peaks with early industrial development before improving as the economy becomes more 'developed', such arguments only ever marshal evidence at the level of individual nation-state economies, rather than the global capitalist system as whole. Yet, continued economic growth paired with de-industrialisation in the global North has only been possible because these polluting industries have moved elsewhere, notably East and South Asia, where they are also likely to be more polluting than they were before. At the global scale, therefore, the theory is patently false.

Moreover, even to the extent that pollution controls improve alongside economic development in individual countries, there is no precedent for reduction in greenhouse gas (GHG) emissions while economic growth continues (Ockwell 2008). Similarly, the vaunted 'dematerialisation' of economic growth and/or its 'decoupling' from increased resource use is simply a 'myth' (e.g. Huws 1999, Jackson 2009). Predictably, therefore, industrialised farming has also become a major source of pollution, including greenhouse gases (e.g. IPCC 2007), especially in the form of nitrous oxide (N_2O) from mineral fertiliser and methane from livestock, which are respectively around 300 and 25 times more potent than CO_2.

In short, while one cannot say definitively what will happen in response to the ongoing attempts at the real subsumption of agricultural nature – and concrete events will, of course, differ from case to case, given the irreducible unknowability and recalcitrance of *particular* socio-natures – current trends augur only the further exacerbation of the present agrarian (and environmental) crisis with, at best, temporary settlements that simply shift the costs on to others, elsewhere and/or in the future.

3.5 Conclusion

Interest in the economics of science has mushroomed in recent years with the growth of various 'new economy' policy discourses that place knowledge, including science, at the heart of economic growth. The latest incarnation of these discourses is the EU's (and OECD's) vision of a KBBE. This builds on prior ideas of the knowledge-based economy, most recently incorporating concerns about the environmental sustainability of economic growth in the promise of a sustainable and equitable new economy characterised by everlasting and ecologically friendly economic growth ('sustainable capitalism') based on the limitless, recyclable and autopoietic characteristics of knowledge and nature as major sources of value ('sustainable capital' per Birch *et al.* 2010). We have seen how this vision is fundamentally false in both respects.

This policy discourse is not merely vacuous, however, but is actively shaping the actual construction of a 'new' economy but one that is radically different to this vision. Furthermore, the emerging actual KBBE is littered with problems and 'overflowings', many of which are manifestations of the fundamental contradiction of the capitalist mode of production (CMP) between use-value and exchange-value, and that exacerbate and condition the agrarian crisis they are supposed to address. These developments are, in turn, crucial aspects of the real, contemporary political economy of science, not least because this imaginary policy accords techno-science (including agricultural science) a crucial, indeed central, role in this new economy but on the basis of a highly restricted (and ill-founded) definition of 'science'. This is thus transforming how, which, where and by how much science is funded, in ways that, in turn, both co-produce the social context of further science (and innovation) policy and are catalysing the emergence of further problems and 'overflowings'. It is also conditioning the sidelining and corrosion of alternative visions – for a different 'new' and sustainable 'KBBE' and a different science – with possibly greater *realistic* attention to the ecological and cultural political economic reality from and upon which we must build.

Failures (and controversies and scandals) must thus be expected for the KBBE narrative. But given the enormous concentration of political power in global agribusiness TNCs and the structural conditions underpinning (the imaginaries of) privatised, proprietary and commercial research, such Macmillanesque 'events' can be expected only to result in further tweaks rather than the fundamental reorientation that is needed, at least in the absence of political mobilisation in support of alternative futures.[15] But what is this 'structural' conjunction underpinning the (unsustainable) neoliberal order of the political economy of knowledge (including science) and why does it need proprietary research, a new enclosure movement? It is to these questions that we now turn, regarding the global rise of intellectual property rights.

Further reading

Birch, K. and V. Mykhnenko (eds) (2010) *The Rise and Fall of Neoliberalism*, London: Zed Books.

Kloppenburg, J. (2004) *First the Seed: The Political Economy of Plant Biotechnology*, 2nd edition, Madison, WI: University of Wisconsin Press.

Lave, R., P. Mirowski and S. Randalls (eds) (2010) Special Issue on 'STS and Neoliberal Science', *Social Studies of Science* 40(5).

Magdoff, F., J.B. Foster and F. Buttel (2000) *Hungry for Profit*, New York: Monthly Review Books.

Weis, T. (2007) *The Global Food Economy: The Battle for the Future of Food and Farming*, London: Zed Books.

4

INTELLECTUAL PROPERTY RIGHTS AND THE GLOBAL COMMODIFICATION OF KNOWLEDGE[1]

4.1 Introduction

A key element of current attempts to commercialise and commodify science is the increasing strength and prevalence of intellectual property rights (IPRs) in standard research practices, both in industry and in academia. For the appropriation of scientific knowledge for private economic gain depends entirely upon legally enforceable property rights, especially in the case of codified knowledge that approximates the characteristics of a 'public good', and so is almost impossible to withhold from non-paying customers without the support of the law. With a social goal of maximising knowledge production, however, the cumulative nature of these processes entails that a balance must be struck between sufficient incentives for the production of knowledge and reciprocal incentives or obligations for its public dissemination and disclosure. Accordingly, IPRs may be more properly described as 'intellectual monopoly privileges' (IMPs) (Drahos and Braithwaite 2002), i.e. temporary monopolies over intangible assets granted by a state in exchange for various forms of public disclosure and access. Accordingly, we will use the latter terminology in this chapter.

The strengthening of IMPs associated with the neoliberal KBE, however, has been a global phenomenon, under the aegis of the Trade-Related Intellectual Property (TRIPs) agreement of the World Trade Organisation (WTO). But the signing of this international treaty in its current form is fundamentally baffling, for even its economic apologists admit it is in the economic interests of only a handful of developed economies (at least 'in the short term'), or rather the transnational corporations (TNCs) domiciled in these countries, particularly the US.[2] How then did it come to be signed into international law by all the member states of the WTO when so many stood only to lose economically from doing so?

It is clear that the pre-eminent political agents for TRIPs, at least as regards patents, were the pharmaceutical TNCs, in particular those based in the US (henceforth 'big pharma'). Big pharma thus not only entirely co-opted the machinery of the US state in international trade negotiations, but did so to the extent that the US Trade Representative (USTR) was prepared and able to overcome the strong objections of most other signatories of the agreement. But why was big pharma thrust to the position of exceptional political enablement?

The question matters not just as a theoretical conundrum. Rather, as we shall discuss in this chapter, explaining this exceptional development – a veritable and unprecedented global corporate coup – discloses the structural conjuncture of the global capitalist political economy that has conditioned the entire neoliberal globalisation and KBE project, as well as highlighting the central role played by the commercialisation of science (especially in the US and in the life sciences) in this process.[3] In particular, this chapter argues that, while there are numerous well-sourced accounts of how TRIPs was effectively drafted and implemented by US big pharma, this kind of agential history overlooks some hugely important contextual issues, without which it becomes hard to understand how big pharma achieved such a stranglehold over the machinery of US state trade diplomacy.[4]

This opening up to context takes two stages. First, a critical history of big pharma reveals that global patent reform was not its only political demand at the time of TRIPs regarding which it was overwhelmingly successful. *Domestic* patent reform in the US was also pursued and in these demands it was supported by two other agents that were crucially affected: the fledgling biotechnology sector and the life science departments (above all molecular biology) of leading US universities. For two of the central demands of domestic patent reform were the extension of patentability to biological material and to publicly funded basic research, both of which mattered greatly to these agents. These political demands went hand-in-hand with the formation of a university-industrial (U-I) complex based on biotechnology, in a mutually complementary dialectic: the greater the success in domestic patent reform, the greater the coherence and power of the U-I complex and vice versa.

Both these political demands and the formation of the U-I complex were overwhelmingly successful. The political successes, however, cannot be explained just in terms of the coherence of the interests of the three parties to the complex, for this reckons without the countervailing interests resisting its formation. And in each case, substantial controversy and political conflict accompanied the proposed changes. Yet all three parties have experienced extraordinary commercial successes and on remarkably similar timelines, from a watershed of 1980. There is thus a striking parallel between the three developments matched by their participation in each case in the U-I complex. The question thus arises: 'what explains these parallels?'

This takes us to the second stage of unfolding the context of TRIPs with a turn to the particular structural context of these developments, in order to explain the conditions that enabled the emergence of such exceptional political agency. Such a structural analysis reveals the various demands regarding patents to be perfectly consonant with the structural imperative for the global primitive accumulation of knowledge along three dimensions: extensive spread of capital into new areas of the globe by TRIPs; intensive spread within already-capitalist societies into new technological capabilities by biotechnology; and intensive spread into the social relations of the production of knowledge through commercialisation of the university. Thus, while the political agency behind TRIPs was big pharma, its capacity to take over the state's agenda for international trade diplomacy regarding patents was built upon the domestic political success of a much broader coalition, a U-I complex in the life sciences, and this success in turn was possible due to its unique complementarity with structural imperatives; an open-ended and mutually confirming process entirely dependent upon active but structured and structuring political agency.

The chapter proceeds as follows. The first section presents brief critical histories of each element, showing how their individual economic interests led to the pursuit of business relations with the other two parties and to the complementary demands for patent reform to facilitate the formation of the U-I complex. The necessary conditions for the formation of a highly integrated political coalition were thus in place. The second section then briefly outlines the controversial nature of each of the patent demands. This is followed by consideration of their total success and overnight change in fortunes from 1980. By placing these histories in a novel juxtaposition, they are given new and important significance. This is particularly so once set in the context of the final section, where comparison of the demands with the structural context is shown to explain the extraordinary success in each case. The chapter concludes with consideration of developments in global intellectual property law and practice since the signing of TRIPs illustrating the ongoing neoliberalisation of knowledge production.

4.2 Critical history 1: Transnational pharmaceutical corporations

The pharmaceutical TNCs at the heart of the TRIPs negotiations and the formation of U-I coalition, such as Pfizer, are one major element of a huge, complex and global industry. While these TNCs are based in a number of developed countries, the majority are domiciled in the US. The rise of the US pharmaceutical industry occurred in particular with the production of penicillin during the Second World War. The development of industrial fermentation processes for production of antibiotics, and the establishment of cartels, served to propel the American companies to dominance (e.g. Dutfield 2003, Liebenau 1984, Swann 1988). By the 1970s, pharmaceuticals was thus a major, profitable sector of the US economy, with profits throughout the decade at 10 per cent, approximately twice that of the average for the economy as a whole (see Figure 4.1 on p. 101).

The nature of pharmaceuticals, as a 'knowledge-intensive' business, also makes it particularly suitable to transnational activity, because the most important advantages of such operations over their local competitors will typically be technological or knowledge-based.[5] Accordingly, by the 1970s it was already a thoroughly multinational industry.[6] The transnational spread of research and development (R&D) of drugs upon which the major players depend was also growing though predominantly in other developed countries (Kuemmerle 1999).[7]

One of the stimuli for moving such operations overseas, however, was the visibility from the 1970s of the increasing costs of R&D, a trend continuing to the present (Drahos and Braithwaite 2002: 59). While a drug in the 1960s averaged $5 million to develop, by the mid-1970s this had grown to $25 million and today the figure is considerably higher again, probably over $100 million.[8] A major cause of this was the increasing regulation of drugs (particularly following the thalidomide scandal) in the major developed country markets (Dutfield 2003: 97). However, increasing costs were also compounded by a diminishing rate of discovery from the traditional processes of screening of synthesised chemicals (Drahos and Braithwaite 2002: 59, Dutfield 2003: 96).

The prospects for the US industry seemed to be further jeopardised by the growing competitiveness of other pharmaceutical firms, both the other TNCs based in Western Europe and Japan and a growing generics industry in the developing countries. With the rising costs and competition, therefore, drugs now had to be 'sold worldwide, since no company can fully exploit a patented product, recouping the R&D costs solely in its own home market, even in the two largest national markets, the USA and Japan' (Dutfield 2003: 168, Archibugi and Iammarino 2002). US big pharma was thus seeking new markets to maintain profits, cheaper production and new R&D processes that would break through the threatened exhaustion of the pipeline.

There was, however, a crucial factor complicating this process, namely the exceptional dependence of the pharmaceutical industry on patents. On the one hand, big pharma sought global markets for realisation of profits commensurate with the rising costs of R&D; but on the other, without patents the growing technological capability of local competition threatened merely to exacerbate the problems of profitability because reverse engineering and imitation was increasingly likely.

There is little doubt that the pharmaceuticals industry 'stands alone' regarding its dependence upon patents.[9] The reasons for such dependence are also relatively straightforward. The pharmaceutical product is very expensive to develop (as we have seen) but a major part of this expense is the regulatory process of clinical trials, which do not produce an advantage built into the product that is easily appropriated.[10] Furthermore, many pharmaceuticals are relatively easy to reverse engineer (Richards 2004: 141). In particular, by the late 1970s local competitors in the developing countries were gaining the necessary technological capability for such reverse engineering. This was compounded by patent reform in these countries, specifically aimed at fostering such a national industry by weakening drug patents. The resulting fall in market share of big pharma in these countries, set against the (long-term, as we shall see) need for expansion of markets into these very countries to recoup the growing costs of R&D, thus made patents *even more* important for the profitability of the TNCs.

Yet, we must not overstate the importance of these developing country markets for these TNCs. The US still today remains the most important market, and sales in the developed world continue to dominate massively those in the developing world. Patent reform within the US itself was thus also of great importance to big pharma. In the context of a generalised anxiety in government circles about the declining competitiveness of American industry, this profitable TNC industry was well placed to receive a favourable audience from government over its concerns. Yet, the dwindling pipeline could have easily undermined this position if it had allowed pharmaceuticals to be plausibly cast as a mature and declining industry, for the traditional free-trade sentiments on Capitol Hill would have baulked at supporting protectionist measures.[11]

Big pharma, however, could claim two particular competitive advantages that greatly strengthened its political hand: its central association with the 'technology of the future' and with a US asset of unassailable superiority. In short, in addressing its pipeline problems, big pharma could call on its culturally credible connection, respectively, with biotechnology and with the source of this 'technological saviour', American university basic research in the life sciences, particularly molecular biology. The huge importance of these two factors was noted at the time by Kenney (1986: 33), 'In nearly every current statement of the present economic malaise, the university is looked upon as the source of new technologies which are to spark sustained long-term recovery.' Biotechnology was the primary example. The credibility of this claimed link, however, and the subsequent formation of a complex of big pharma and these university departments, mediated by biotechnology, depended upon the reception of big pharma's advances from these other two parties. It is to these that we now turn.

4.3 Critical history 2: Biotechnology

Faced with the prospect of a declining output of new drugs, by the late 1970s pharmaceutical (and chemical) firms were looking for new alternatives. With the underlying similarities in biomedical subject matter and the historical experience of the enormous success of applying organic chemistry in industry, the emerging research in biotechnology was quickly latched on to as the 'obvious' next step, the technology that would resolve the US's (and the globe's) economic woes (Drahos and Braithwaite 2002: 59, Dutfield 2003: 148).

While the term 'biotechnology' originally referred to fermentation (Bud 1992, 1993), the biotechnology arousing such excitement arose from a number of seminal breakthroughs in molecular biology, the first and most important of which was the discovery of recombinant DNA (rDNA) techniques by Cohen and Boyer in 1973. These discoveries emerged in relatively quick succession predominantly in molecular biology departments in leading US universities and it was quickly apparent that they were 'transform[ing] radically the knowledge base and the opportunities for innovations in biotechnology' (Orsenigo 1989: 36).

The historical connection of big pharma with universities meant that it was the first industry to recognise the significance of these developments (Orsenigo 1989: 85). Molecular biology, however, was not a department in the life sciences with strong historical links with industry. Nevertheless, it was a discipline that was particularly compatible with the demands of business for profitable biological commodities because it had been developed from the outset as a biological discipline that would contribute to social control (Kay 1993, 1998).

While its programme was not initially concerned primarily with commercial applications, its uniquely reductionistic viewpoint, reducing 'life' to the interaction of a number of molecules, adapted easily to the imperatives of the production of appropriable, and so isolatable, biological commodities (Kay 1998: 35, Kenney 1986: 9). The social structure of the discipline also facilitated an easy fit into relations with industry: the use of hi-tech apparatus along with the large budgets and responsiveness to funding imperatives this entailed, as well as the hierarchical organisation of the laboratories, all meant the discipline had significant experience of working in an appropriate environment (Kenney 1986: 12).

The 'new' biotechnology was thus arising from autonomous developments at the level of 'basic research' in molecular biology. Indeed, those interested in developing such biotechnology found it to be exceptional as a new technology in its dependence on 'basic science' capabilities and the publicly funded research conducted in universities (e.g. Orsenigo 1989: 2, Rosenberg and Nelson 1994: 343). It followed that the major barrier to entry into, or rather development of, a biotechnology industry was access to the relevant scientific expertise located in universities and the need for the creation of links between industry and these individuals and their departments (Orsenigo 1989: 40, 49). In short, for this biotechnology to be successfully developed, the creation of a university-industry complex was necessary.

This did not mean that biotechnology entrepreneurs would necessarily welcome the advances of big pharma. Not every start-up in a new technology wants dealings with the existing big players because, if it can successfully develop the technology without such assistance, efforts to provide it will be legitimately treated as attempts to protect existing market positions. But this was not the case for biotechnology. For the very centrality of the basic science meant that development of successful commercial products was neither founded on a relatively unsophisticated scientific base nor a straightforwardly appropriable process, taking place at the level of technological tinkering. Unlike computers, biotech could not be developed by enthusiastic individuals in their garages. Biotechnology thus actively sought large external investment, without which it was impossible to launch, and big pharma was an obvious source of such financial support. Furthermore, with biotechnology's strengths lying in research, the established capabilities of these large firms in development and marketing promised a synergy of business operations to both parties (Kenney 1986: 207, 1998: 137). In soliciting biotechnology as a potential solution to its pipeline problems, therefore, big pharma found a willing partner.

The particular political goal of big pharma, recall, was patent reform. It was crucial for this political strengthening, therefore, that the interests of biotechnology entrepreneurs were compatible with big pharma's not merely economically but also regarding this domestic political demand. Biotech, however, also clearly stood to benefit from patent reform. This again arises from the exceptional central role of 'basic science', because such science, as high-level knowledge relatively disembodied from technical know-how, is relatively easily appropriated by competition. Biotechnology start-ups are thus exceptionally secretive (Kenney 1986: 177). They are also high risk and extremely research intensive, with R&D expenditures commonly as high as 50 per cent of sales, so that such threats of imitation jeopardise the whole business plan (Dutfield 2003: 153). Furthermore, the level of activity meant that such technological lead times as did exist were quickly shortened considerably. While a business can always resort to trade secrets in the absence of patents, so that it is difficult to argue that patents were 'necessary' for the development of biotech, the added security of patents was evidently enormously attractive to biotech start-ups. Big pharma's interest in patents was thus matched by a similar interest from biotech.

For patenting to be possible in biotech, however, two particular changes in the law seemed to be necessary. First, it was unclear that biological materials were patentable at all in that they could fall foul of restrictions on patenting scientific discovery rather than invention.[12] Second, given the location of this biotechnological research in university departments, the illegality of the patenting, and hence private appropriation, of the results of publicly funded basic science research in universities was a major problem.[13] Without these changes, providing the 'legal certainty' required for assurance to potential investment, biotechnology and the formation of the U-I complex necessary for its development was significantly less assured. For their own economic interests, therefore, both big pharma and biotech actively sought such domestic reform. Its success, however, depended on whether one final player, the relevant life science university departments, would accept the radical transformation in norms that such legislative reform would institute.

4.4 Critical history 3: The privatisation of US university research

Federal funding of science ballooned during the Second World War (e.g. Mirowski and Sent 2002a, Mowery *et al.* 2004). The massive 'success' of converting basic research into crucial technologies, such as the atom bomb, and then the scare of its cold war rival's launching of Sputnik, meant that this funding continued and indeed increased consistently in the following decades. At the beginning of the 1970s, however, the percentage of federal funding of university research began to fall. By 1980, dire economic circumstances were increasing the calls both for slashing the science budgets of what was easily branded a cosseted and politically radical elite and for the universities to prove their contribution to the faltering national economy. With the new Reagan administration duly obliging with threats of such cuts and with high inflation eroding the real value of what funds were in fact delivered, the universities found themselves compelled 'to compete increasingly for external dollars that were tied to market-related research' (Slaughter and Leslie 1997: 8).[14] Biotechnology was unsurprisingly the focus of this move because the most striking scientific breakthroughs of the period were occurring in molecular biology. Furthermore, the percentage of federal research funds devoted to the life sciences was large and growing, so that research in this field was plausibly presented as a national competitive asset for commercial exploitation.[15]

Among biology faculty, while there were mixed reactions regarding the embrace of commerce, there was widespread acknowledgement of the possibilities of funding their research through the establishment of such links. From a purely academic perspective of seeking to preserve the funds necessary for their research programmes, therefore, relations with industry were often welcomed. But for such links to be established, patents over research findings were often needed in order to attract investors with the promise of exclusive ownership of the resulting product. Patents also offered the potential of direct licensing income for the departments, again an attractive prospect given the financial constraints of the time.

Patents, however, were particularly important for such technology transfer given the nature of biotechnology (Williams 2000), so that what were portrayed as general, abstract problems were a particular issue for the kind of knowledge that *biotechnology* happens to be. Biotechnology straddles the conventional distinction between 'basic' and 'applied' research, thereby problematising it and stretching economic arrangements premised upon it to their limit.[16] Given this liminal nature, the development of basic research programmes will in many cases be the development of a technology. Yet, this process may be both difficult and very expensive, with the proto-business plan suggested by the original basic research result being only a 'proof of concept'. In the climate of the time, there was little chance that public funds would be available for these projects, but for such investment to be forthcoming from the private sector, the resulting product had to offer the prospects of a profitable commodity.

The simultaneously basic/applied nature of biotechnology, however, presents the business risk that, while development of the actual technology will be very expensive, once successful, public accessibility of the original business idea will render market entrance relatively easy, not least through facilitating reverse engineering. In these circumstances, private enterprise will only take up these risks on the guarantee of some exclusivity in order to be able to recoup and profit from their sizeable investment. But it follows also that patents are necessary for the continuation of such research in universities at all. In the absence of an artificially *created* scarcity of knowledge as produced by intellectual property rights (May and Sell 2006), the traditional 'open science' publication of such knowledge will result in it being stillborn, as a published business plan for an enterprise the particular competitive advantage of which is its *private and secret* knowledge. The only alternative is not to disclose at all, but this is to exclude 'basic scientists' from the game altogether, which takes place instead purely at the level of trade secrets. And, indeed, a 'major factor' in the entry of the university into the biotechnology business was 'fear of losing lead researchers' to pursue their research in private enterprise (Orsenigo 1989: 83).

It must be emphasised that the scenario discussed above is a limit case and that the traditionally recognised means of the appropriating knowledge in the transformation of scientific research into profitable technologies – where lead times and tacit knowledge provide the competitive advantage, so that the avenues of open science remain perfectly acceptable, if not preferable – are by no means obsolete, even in the case of biotechnology (Colyvas *et al.* 2002). Nevertheless, such limit cases were more likely than ever for biotech, and especially in its early days.

University administrators thus began to lobby for this change (Slaughter and Rhoades 2004: 20). Furthermore, the typical argument for the policy in government circles was just such a case of 'university inventions that were embryonic pharmaceuticals' but which would not be developed without a patent (Colyvas *et al.* 2002: 62, Mowery *et al.* 2004: 177, Nelson 2001: 15). The economic *and political* complementarities between the university and big pharma on this score were thus explicit.

The patentability of public research, however, was not the only patent reform in which the universities and life science departments had an interest. For they also supported clarification (if not change) of the patentability of biological materials. Indeed, as Kenney notes (1986: 257), it was the universities who had *most* interest in such reforms. While such a move was certainly important to biotech start-ups and pharmaceutical TNCs alike, they could rely on trade secrets to some extent in the normal course of pursuing profitable business. Yet such action is unavailable to universities for 'the university is not constituted as an institution that sells products in the market. If the university cannot patent inventions it would have nothing to sell because the university cannot use its inventions to become a company' (ibid.).

4.5 Mutual complementarities

For each of the three parties necessary for the successful creation of a university-industrial complex of biotechnology, we have found their specific and autonomous economic interests to be complementary. But this also gave rise to a mutual compatibility and interdependence of *political* needs, because the successful creation of such a U-I complex was conditioned by the necessity of reform to the patent system to allow for such relations to be established and such reform had to be supported by all parties. In each case, this single-issue political support was readily forthcoming. As such, the successful formation of such a complex was not merely the creation of an economic sector but also of a political agent of considerable breadth and coherence focused on domestic patent reform. But, as regards TRIPs, it was also the case that the successful emergence of this patent coalition transformed the domestic political landscape, greatly strengthening big pharma's political leverage at the level of the US state machinery to pursue a regime of strong, *global* patent rights that would cover *pharmaceuticals and biological materials* in particular.

We have already seen that TRIPs was a highly controversial development. Perhaps we might expect, then, that those reforms strengthening big pharma in its demands for TRIPs would be relatively uncontroversial, for the success of a demand besieged on all sides by controversy would be, *prima facie*, unlikely in the extreme. Yet the other patent reform demands were also highly controversial and vulnerable in two ways: first, regarding explicit and vociferous political opposition to the changes; and, second, regarding the hyped or at least essentially unproven claims of the official economic case for the proposed reforms. Indeed, the economic justification of the proposed reforms not only was unproven at the time, but is increasingly exposed to substantial criticism.[17]

Yet, not only were the reforms successful, but the success of the demands of each of the three agents was overwhelming and simultaneous, from a watershed of 1980. How did this occur? The explanation of this success set against the level of controversy attending it demands recourse to the coincidental political enablement from the socio-economic structure. For considerations of structural context, structuring this contingent political agency and the construction of *new* political agents, alone can explain the dramatic turnaround in the fortunes of the various parties. Before turning to this, though, let us consider a brief summary of the controversies attending the various dimensions of the proposed patent reform in order to show the level of the objections and obstacles that had to be overcome.

4.6 Parallel developments: Controversial demands

First, as we have mentioned above, TRIPs was the result of a straightforward political strong-arming in which the substantial objections of numerous developing countries, particularly India and Brazil regarding the potentially penal effects on economic development of redrafting property rights in favour of TNCs, were overridden (e.g. Drahos and Braithwaite 2002, Sell 2003, Shiva 2001). Second, as regards biotechnology, from the time of the initial rDNA successes, there was considerable debate about the regulation of these technologies with national and local government threatening serious limitation of the use of rDNA techniques for reasons of safety (e.g. Kenney 1986, Krimsky 2003, Wright 1994, 1998). Finally, the increased connections between academe and industry that biotech involved were and continue to be contested, with numerous scholars and commentators expressing serious misgivings over the effect of such commercial ties along a number of dimensions: for example, on the scientific 'neutrality' and status of the university as a knowledge-producing institution; on research agendas; and on the likelihood of greater obstructions to the free flow of scientific findings.[18]

But the cases for these reforms were not particularly strong either, being couched in the familiar terms (discussed above in Chapters 1 and 2) of the 'linear model' (of academic 'basic science' giving rise to applied science or technology, which then produces economic growth) and the 'public good' argument of knowledge production. Patents, by creating property rights over such knowledge products, are one way to resolve this problem (David 1993). Yet, we have seen that both elements of this argument, linear model and market failure, have been seriously criticised, both generally and in respect of biotech. In fact, most industries do not depend greatly upon 'basic science', nor upon patents for innovation. Nor do they depend upon the patented results of basic research for its dissemination and commercialisation.[19] Channels of open science are generally deemed much more important, suggesting a much more complex role for the university as regards its contribution to economic growth than the policy rhetoric allowed. Furthermore, in direct opposition to much of the discussion of the time, far from being an easy development and a safe investment, the 'obvious' next step, biotech was a highly uncertain and risky technology (Orsenigo 1989).[20] Explanation of how and why it was treated as such despite these problems is thus required.

Finally, big pharma's plea for strong patents in developing countries as being necessary for its profitability as a business model is belied by the fact that, still in 2003, 91 per cent of sales and 98.8 per cent of R&D expenditure was associated with developed countries, and in particular the US, where patents were already strong before TRIPs.[21] Indeed, the CEO of Pfizer and architect of TRIPs himself admitted that losing market share in developing countries in the 1970s had very little impact on profitability (Drahos and Braithwaite 2002). The astronomical estimates of the costs of developing new drugs have also been seriously questioned, undermining the central plank in the argument.[22] To be sure, drug development is a very expensive business. But realistic estimates of costs set against the very small contribution of developing country markets to big pharma's profits make the argument for TRIPs, *even* (if not *especially*) *in the case of pharmaceuticals*, highly problematic. In all these respects, therefore, the cultural credibility of the arguments for TRIPs themselves must be explained.

4.7 Parallel developments: Extraordinary successes since 1980

Success of TRIPs and pharma

The most astounding success of big pharma is, of course, orchestrating TRIPs itself, an unprecedented agreement providing strong patent rights on a global scale and in particular to the industries (of pharma and biotech) for which such provision is most controversial. Indeed, such was the success that a key player in the big pharma negotiations, Jacques Gorlin, was moved to state that the agreement was '95 per cent' of what they wanted; the ominous remaining 5 per cent referring to compromise on compulsory licensing and a limited provision for a public health emergency carve-out (Drahos and Braithwaite 2002, Sell 2010).

But pharma has also prospered spectacularly in the last 25 years, from the 'watershed' of 1980 (Angell 2005: 3) when profits in the industry took off (see Figure 4.1). The industry's profits were a healthy double of the average throughout

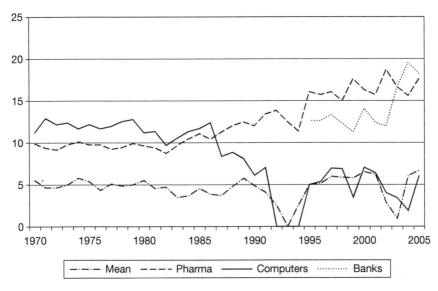

FIGURE 4.1 Profits/revenues for top 500 US companies, 1970–2005 (%)

Source: Fortune 500 data, compiled by author.

the 1970s, but from 1980 this ratio has grown to an average 2.4 for the 1980s and 3.3 for the 1990s.[23] In 2002, in the midst of recession, profits for big pharma ($35.9 billion) were greater than profits for all of the other 490 companies of the US *Fortune 500* combined ($33.7 billion) (Public Citizen 2003b).

Nor has this increase in profitability been the commensurate 'reward' for increased innovation in the industry, which has been falling throughout the period. Of the new drugs approved in the US between 1989–2000 only 15 per cent were of the highly innovative class that provide a significant improvement over existing drugs (NIHCM 2002: 9). Rising prices thus reflect not the remuneration of increased innovation but the costs and successes of numerous other tactics, particularly marketing and legal 'strategy'. As regards the former, official figures on expenditure by big pharma on marketing are now consistently significantly greater than R&D, and this despite the fact that much marketing to the medical profession is reclassified as 'education' (Angell 2005: Chapter 8).[24] As regards the latter, the most important tactic for maintaining profits is to prolong patents. For instance, firms can get 'a new patent and Food and Drug Administration (FDA) approval for a trivial variation of their blockbuster and promot[e] it as an "improved" version of the original' (Angell 2005: 183).[25] Patents are thus routinely prolonged for years after their theoretical expiry date.

A third major means by which big pharma maintains its profits is by political lobbying on a massive scale in a runaway positive feedback loop: the profits afford further penetration of pharma lobby activities into government, which in turn produces legislative action to further bolster its profitability. This is particularly marked in the US. There, the industry already has influential ears in government at all levels with many high profile contacts and a 'revolving door' between government and lobbying (Public Citizen 2001a, 2002, 2003a, 2004). But the huge numbers of professional lobbyists it employs at great expense (reaching nearly 1,000 lobbyists at over $140 million in 2003) redoubles its access to the political machinery. Evidently, such expenditure must be both affordable and value for money. The profitability of the industry assures the former, but the latter is also amply in evidence with legislative decisions in its favour too numerous to list exhaustively, including the extension of monopolies and tax breaks (Angell 2005).

This almost unchallengeable power is ripe for abuse and evidence of such malpractice is rife and growing. Given the seriousness of their products for human health, perhaps the most important abuse is the rigging and bias of clinical trials; something that can be done 'in a dozen ways' and is now 'rampant' (Angell 2005: 95). The result has been a number of highly excoriating reviews in the leading medical journals of the *New England Journal of Medicine*, *JAMA* and the *BMJ*, but few active political steps at reform through the entire Bush presidency and now into the Obama administration (Bekelman *et al.* 2003, Bodenheimer 2000, *British Medical Journal* 2003, respectively). This is to list only the most egregious example, because the purpose of this volume is not primarily to expose or condemn the pharmaceutical industry.[26] Yet, these abuses have yet to be brought in check, providing further evidence of big pharma's extraordinarily privileged political position.

Biotech success

Turning next to biotechnology, the most remarkable fact about its development is that, despite the strong uncertainty associated with it, discussed above, 'attempts to exploit it commercially *immediately followed* the scientific discoveries' (Orsenigo 1989: 2, emphasis added). The trigger was the success, in autumn 1977, of expression of the somatostatin gene in *E. coli*, which led to take off in TNC and venture capital investment in small biotech and genetic engineering firms (Wright 1998: 93).

At this time, the regulation of rDNA techniques remained exceptionally high profile in formal political debate, and, indeed, seemed very likely (ibid.: 91). Yet, in November 1977, the US Senate Subcommittee on Science, Technology and Space saluted the 'extraordinary progress towards the construction of organisms that make therapeutically useful hormones' (quoted in ibid.). The statement was merely a harbinger of a much more profound shift in the debate, in which 'the "real" risk was now defined as that of losing out on a novel field with immense commercial impact' (ibid.). This culminated in regulations being relaxed twice in 1980, and again in 1981 and 1983, in effect totally dismantling controls on rDNA experiments (Kenney 1986: 26–32, Wright 1998: 97).

In this context, and providing the background to these changes, there was a 'gold rush' of IPOs (initial public offerings of shares) in the first biotech companies to be floated on the stock market.[27] For example, in 1980 Genentech's IPO was the fastest rise of any stock 'in the history of the New York capital market' (Bud 1998: 5, Kenney 1986: 156, 1998: 136). While a tailing off in numbers of IPOs followed, up to 1983, this was in the context of a deep recession and hence a weakness of market conditions for IPOs (Kenney 1986: 137). When the economy recovered in the latter half of the decade, so too did IPOs. Similarly, that there was a shake out of the sector is not nearly as remarkable, given the background economic conditions of the time, as the *lightness* of this effect on biotech. 'Business failures were at all-time highs, yet biotechnology expanded into the recession' (ibid.: 175).

The result of all this business interest is a significant biotechnology sector: in 1998, there were 250 public and approximately 1,000 private companies (Arundel 2000, Thackray 1998). Yet, this growth in the sector has occurred in the context of relatively few successes. Only 54 biotech-derived therapeutics and vaccines had been approved in the US by the late 1990s, including a rush in 1998 (Senker 2000: 58). And while many of these products are 'of significance for health and for agriculture, their economic impact is far more limited than the impact of computerisation in 1970' (Arundel 2000: 84). In short, biotech remains, over 20 years after initial investment excitement, a business that, while undeniably up-and-running, is built largely on expectation.

US patent reform since 1980

Since 1980, there has been a host of patent and drug legislation, all of which has favoured the U-I complex (Slaughter and Rhoades 2002). Each of these acts of legislation provided a major boost to the U-I complex, in particular the Bayh-Dole Act, which allowed patenting of university research (e.g. *The Economist* 2005). We will discuss university patenting in more detail below. Here we focus on a number of other developments in patenting more generally. The legal development of greatest importance for biotechnology in this regard was, in fact, a judicial and not a legislative change: the judgment of the Supreme Court in *Diamond v. Chakrabarty* [1980], which provided an indubitable legal authority to the patentability of biotechnological commodities.[28]

A second change of enormous significance was the establishment in 1982 of a separate court circuit dedicated to patent litigation. A specialised court for patent litigation is staffed by patent lawyers who are inevitably more attuned to patent issues than those of other economic interests or imperatives, such as competition policy. The result has been a significant strengthening of patentees in such litigation, and hence of patents generally (Jaffe 2000: 549). For instance, while nearly two-thirds of adjudicated patents were found invalid by the courts between 1921–73, between 1982–85 this fell to 44 per cent (Boyle 1996: 134). Waldfogel (1998) notes that the chances of successful defence of a patent have risen to 84 per cent in cases lasting 3 months and 61 per cent of those lasting one year by the mid-1990s and without significant change since then.

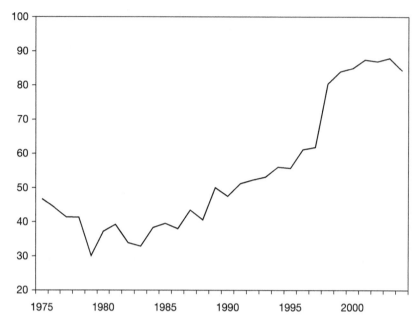

FIGURE 4.2 US patents granted per year (thousands), 1975–2005

Source: United States Patent and Trademark Office.

This development has been complemented by a widely observed weakening in the standards of both the novelty/non-obviousness and utility tests by the patent authorities (e.g. Barton 2000, Cohen and Merrill 2003, Jaffe 2000, Jaffe and Lerner 2007, Merges 1999). Whatever the cause of this, patents are now not only stronger, but also more easily granted in the first place. It is thus predictable that there has been a massive growth of patenting in the US, again since the early 1980s.[29] Between 1985 and 1999, the number of new patents granted per year doubled (see Figure 4.2) (Gallini 2002: 131, Kortum and Lerner 1999). The growth of patenting is particularly noticeable in biotechnology. For instance, the number of patents in the patent class of 'molecular biology and microbiology' increased over ten-fold between 1985 and 1998.[30]

University patenting and privatisation

We discussed above how patenting has ballooned since 1980, but this is particularly marked for universities. Thus, while patenting increased through the 1970s, it exploded in the early 1980s: university patents more than doubled between 1979–84, again in 1984–89 and again in 1989–97, reaching a peak in 1999 (see Table 4.1). Similarly, between 1980 and 1997, university patents per dollar more than tripled (Jaffe 2000: 541). This growth was concentrated in the largest 100 universities, which tripled annual patent output from 1984–94 (Gallini 2002: 131).

TABLE 4.1 Utility patents issued to US
universities and colleges,
1969–2004 (year of issue)

Year	Number of patents
1969	188
1974	249
1979	264
1984	551
1989	1,228
1994	1,780
1999	3,363
2004	3,057

Source: Mowery *et al.* (2001) and United States
Patent and Trademark Office (2010).

The growth of patenting, however, has not been limited to these leading institutions, but has spread across the entire education sector. Thus, in 1965, 30 academic institutions received patents, growing to 150 in 1991 and over 400 in 1997 (Jaffe 2000: 541). Similarly, there was an eight-fold increase in university technology transfer offices between 1980 and 1995 (Cohen *et al.* 2002: 4). As the universities had hoped, this has been lucrative, though only for a few universities.[31] Licensing revenues grew to $222 million in 1991 before trebling to $698 million in 1997 (Mowery *et al.* 2001).

As is the case in the economy generally, the growth of such university patenting has also been 'disproportionately concentrated in technological classes related to health sciences' (Jaffe 2000: 541).[32] At Columbia and Stanford Universities, both major protagonists and beneficiaries of the change, by 1995 biomedical patents accounted for more than 80 per cent of their substantial licensing revenues (Mowery *et al.* 2001: 107).

Increased patenting is merely one element of a broader change in the funding of basic science in the university through the proliferation of U–I relations. It must be stressed that this was not a black-and-white change from a golden age of public sector 'purity' to a dark age of private sector 'corruption', a point rightly emphasised by Mirowski and Sent (2008). Industrial links were strong in the US university system before the Second World War and remained so even at the height of the cold war 'open science' regime (Kleinman 2010). Nevertheless, this continuity has been accompanied by a fundamental discontinuity, namely 'the extent of these ties and the intensity of their effects' (Wright 1998: 94).

The overall funding of academic science shifted considerably from public to private sources in the period from 1970, with the percentage of federal funds falling through the 1980s (though actual amounts grew consistently) (see Figure 4.3).[33] On the other hand, there was a growing, if still small, percentage of industry

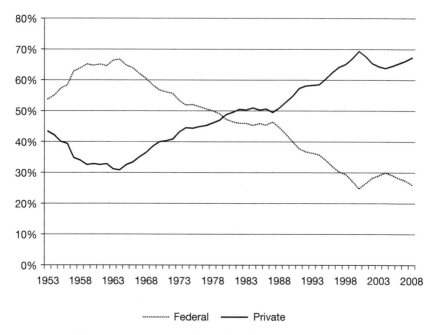

FIGURE 4.3 Federal versus private funding as % of total R&D budget

Source: National Science Foundation (2006, 2010).

sponsorship, from a low of 2.6 per cent in 1970, growing slowly to 3.9 per cent in 1980, then mushrooming to approximately 7.0 per cent in 1990 (a nearly five-fold increase in nominal dollars), where it remained for the rest of the decade (Cohen *et al.* 2002, Narin *et al.* 1997, National Science Foundation 2006). Once again, however, this scenario was intensified in biotech, where industry funding was already over one third in the mid-1980s (Blumenthal *et al.* 1986a, 1986b).

Orsenigo (1989: 77) thus notes the 'explosive rate of growth' of U-I relationships in biotech, the 'most striking feature' of which is 'the direct involvement of scientific institutions and industrial scientists on the one hand, and the systematic, targeted nature of the ties on the other'. Most of these relationships were with 'only a limited number of companies (mainly chemical or pharmaceutical corporations)' (ibid.: 81). Similarly, Krimsky (2003: 31) calls the growth in U-I relations in the 1980s 'unprecedented' and 'a fact specifically evident in biotechnology'.

Success of the coalition

In search of means to boost their respective finances, big pharma, biotechnology entrepreneurs and university life science faculty and administrators all committed themselves to the development of biotechnology and the formation of U-I relations necessary for this to occur. We have seen above how both of these have been extremely successful. Thus, we can have little doubt that this biotechnology U-I complex has indeed been created, creating a synergistic mutual dependence between the various parties, in which the huge capabilities of big pharma to develop and market products is matched with the unique research capacities of the biotech firms (Kenney 1998: 140, Ronit 1997).

But the forging of these links has also assured the inextricability of the fates of university life science research and the pharmaceutical industry, providing the latter with a potential trump card in political wrangling for the foreseeable future.[34] The major political success of the coalition was domestic US patent reform, but the success of biotech itself was also utterly dependent on the social acceptance of the privatisation of publicly funded research (Kenney 1998: 135), acceptance that has been politically achieved with bipartisan support in Congress (Slaughter and Rhoades 2002). The formation of a U-I coalition by these agents thus marks a watershed in the domestic political landscape regarding patents.

This is not to argue that pharmaceuticals, biotech and university life sciences are now an indissoluble unity. They remain clearly distinct with their own institutional imperatives and interests. Yet, pharmaceutical research remains overwhelmingly dominant in biotechnology.[35] And the pharmaceutical industry has provided the 'protective niche' necessary for the development of this fundamentally risky and radically novel technology (Arundel 2000: 86). The success of this process is apparent in the fact that even failure (Bud 1998: 4) and continuing controversy, flaring up again from 1996 (Bauer and Gaskell 2002), have had only relatively superficial impact on the hopes surrounding biotechnology.

The role of patents in this success cannot be overstated. For they have been crucial in the management of commercial relations between university and industry upon which biotechnology has been built (Williams 2000: 68 ff.). Nor is this due to commercial interests perverting the development of the technology to its nefarious ends. Rather, as we have seen, given its nature on the basic/applied science boundary, for university research to be possible in this form of biotech, in the context of a capitalist economy, patents (though not patent *reform*) were needed.

The penetration of commercial relations and patenting into these departments, therefore, has not been something *driven* by the pharmaceutical industry so much as thoroughly *exploited* by it, in particular by using the political power this presented it to orchestrate patent reform not only within the US but also strengthening IMPs across the globe in its favour. To repeat, even domestic patent reform was not *necessary* for the development of biotech. Patents were already available to an extent and, in any case, are not always needed for technology transfer even in biotechnology (Colyvas *et al.* 2002: 65). The outcome of these trends, however, is an increasingly proprietary regime of science funding (Orsenigo 1989: 96, Rai and Eisenberg 2003: 291). But this also has significant repercussions for the global political economy. As Williams (2000: 69) puts it:

> The most immediately apparent feature of the socio-economic relationship between patents and the emerging international political economy of

> biotechnology lies in the manner in which the state-sponsored patent system has led to increasing commercial dominance of a number of key sectors in the global economy. . . . The availability of strong patent rights in the field has promoted unprecedented corporate control of the life sciences.

As we shall see in the next section, it is this outcome and the fact that the interests of these agents were such that they would pursue it that together explain their unique political enablement and, in turn, that of big pharma regarding TRIPs.

4.8 Primitive accumulation along three dimensions: The unique enablement of the coalition

There can be little doubt that each of these developments was highly controversial in its own right and yet was extraordinarily successful, on remarkably similar time-scales, in instituting various patent reforms. What, then, can explain these puzzling observations? Not recourse to the exceptional coherence of these agents alone, for this reckons without both the level of opposition they faced and the weakness of each of them as regards their cases. As the comparative failure of contemporaneous attempts to erect a global regulatory architecture for foreign investment and financial flows (in the 'Trade-Related Investment Measures' agreement, or TRIMs) demonstrates, there was no guarantee from the sheer economic size of those in favour that TRIPs would succeed (Sell 2003). Nor can such purely agential analyses explain how, despite the continuity of growing patenting and U-I links in life science departments in the late 1970s, suddenly the social constraints to this process fell away, allowing its explosive growth from 1980 (Mowery *et al.* 2001: 104).

But neither can a purely 'cynical' explanation, which latches on to the fact that the dramatic turnaround in fortunes was associated with the recognition of money being at stake. For such an analysis does not address the crucial question of *why* there was money at stake. Given the highly risky nature of biotechnology, discussed above, it would seem at first to be a particularly *unattractive* investment. That there was money to be made from biotech, therefore, was only *because* of a widespread sense that people were willing, and indeed eager, to invest in it. But then this itself demands explanation and simply pointing to financial greed cannot furnish this.

We must, therefore, explain these developments in terms of the structured agency that constructed TRIPs and forged the coalition that eventually enacted it (Sell 2010). This demands an analysis of the structural context in which these developments occurred, and in particular to the condition of the structure of the global economy, if we are to be able to explain why, *despite* being a risky bet, finance (especially in the US) believed it was a sure-fire winner. In arguing thus, I refer to the argument, discussed above in Chapter 3, of Arrighi (1994, 2003, 2005a, 2005b) *inter alia* that the global economy underwent a fundamental structural change in 1980, triggered by the monetarist revolution of Reagan and Thatcher, from a cycle of strong growth spread across the economy and led by productive business to one in which the economy is dominated by the financial sector ('financialisation').[36]

The fundamental driver of this process is the relentless accumulation of capital, which takes the form of the penetration of the capital relation of waged labour into ever more areas of social life around the globe. This shift was the culmination of a process that had seen the political economic settlement of the post-war period reach its limits in the 1970s in the emergence of a persistent crisis of over-accumulation and then trigger a political crisis that was only averted with a radical reassertion of the power of global capital. This took the form, as it has in previous cycles, of the balance of power between the productive and financial sectors of the economy moving sharply in favour of the latter. The control of finance capital over the economy (as a 'historic bloc'), including via its increasing control of and neoliberal transformation of the state (in what Underhill (2006: 17) calls a 'state-market condominium') in turn allowed it to orchestrate both the radical restructuring of productive enterprise and a round of appropriation into private hands of a new set of resources upon which the global economy will (or at least *has* in past cycles) eventually be able to begin a new cycle of productive growth ('primitive accumulation').[37]

What resources capital seeks to appropriate in these phases is conditional upon the existing extent and nature of the global capitalist economy. In every case, however, capital must expand both outwards, into new social realities inhabiting other areas of the globe, and inwards, intensifying its penetration into the already capitalist societies. The expansion of capital, however, changes societies, and hence is met with resistance. Its expansion in any given round of primitive accumulation is thus both enabled *and constrained* by its contemporaneous extent.

Developments in the global economy are currently characterised by two buzzwords that capture two aspects of neoliberalisation: 'globalisation' and the 'knowledge economy' (as discussed in Chapter 3). On the analysis offered here, these two capture (albeit unwittingly in many analyses) the nature of the present round of primitive accumulation: extensively in the further transformation of non-western economies into capitalist ones and intensively through the transformation of the social relations of the production of knowledge. As regards the latter, the societies at the core of the global economy are already thoroughly capitalist as regards the social relations of production, including as regards the social relations of the production of (what may be broadly designated) 'culture'. For capital to expand yet further into these social realities, therefore, it must transform social relations of production of a kind even further removed from the material reproduction of society. The invocations of a 'knowledge economy' or 'information society', whatever their failings, capture the essential nature of this change: the expansion of the capital relation into the social relations of production *of knowledge* (Caffentzis 2007, De Angelis 2004).

One further dimension needs to be explained. Capital transforms not only social relations but also the material constitution of societies and it does this by the introduction of new technologies that offer profitable manipulations of material reality not hitherto known to be possible. Furthermore, this process is crucial for the continued growth of the economy, for it is novel technological possibilities that afford the profitable transformation of the economy (employing waged labour in the production of surplus value in new conditions) in the Schumpeterian (1976) 'gales of creative destruction' characteristic of capitalist growth. Given that these technologies must be commercially developed in a social milieu, the discovery of such possibilities also opens up a new sphere of socioeconomic life into which capital can expand. Each cycle, therefore, is characterised by a novel technology that percolates through the economy and thoroughly transforms it, and society in the process (see Section V in Volume 2). Furthermore, economic problems start to appear when the possibilities for profiting from such transformation using existing technologies begin to decline. At this time, therefore, in the absence of profitable opportunities from exploitation of presently familiar or 'normal' innovation processes, new technologies will start to attract increasing interest as potential ways to provide new commodities to overcome this stagnation.

The greater the impasse from the exhaustion of the 'normal' development of existing technologies, the greater will be the effort required to develop the 'next' technology. Yet such effort will not be forthcoming until there is sufficient concentration of finance to invest in it. This condition is satisfied during a financialisation phase, but at this time finance's economic power is such that it can also compel the placing in private hands of the necessary resources for the technology's profitable development, including through political reform of the law, redrafting property rights to its benefit. Primitive accumulation thus also takes the form of the development of a new technology that affords the expansion of capital into a wholly new sphere of manipulation of physical reality in the production of profitable commodities.

In the present case, the possibilities for the exploitation of material reality have already been thoroughly consumed in a number of ways: mechanically, electrically, chemically, etc. The new technology must inevitably take its lead from the existing state of scientific knowledge. It is apparent that the discipline that attracts the most attention in this regard is (molecular) biology, and that the technology therefore touted as the 'technology of the future' is biotechnology. We see, therefore, that the hopes pinned on biotechnology arise immediately from the structure of the global economy at that time. The exhaustion of the political economic space since the Second World War and of the possibilities for chemical and electrical products meant that for the global economy to continue its necessary expansion, it needed to find a new and radical technology. But was this biotechnology? And, if so, why?

To answer this question, we must return to consider the twin dimensions of neoliberal primitive accumulation mentioned above: globalisation and the commodification of knowledge. Together, these structural imperatives demanded the 'globalized construction of (knowledge) scarcity' (May 2006: 53). But this is precisely what big pharma sought in TRIPs, strong global patents, and for biological commodities in particular. And we have seen that biotech is exceptional in its dependence upon 'basic' science, thus making it uniquely relevant to the 'knowledge economy'. To be sure, as regards the extensive spread of capital, big pharma's pursuit of TRIPs is merely one element of a broader neoliberal assault, including the International Monetary Fund's (IMF) structural adjustment programmes (Stiglitz 2003) and other WTO trade talks. Nevertheless, TRIPs stands alone as the aspect of this package concerned with the construction of the 'knowledge economy'.

But once we acknowledge that big pharma's political power in the US regarding TRIPs was built upon domestic success in the formation of the U-I complex, we can also see that the complementarities go even deeper in this case. For the prime targets for moves to commodify knowledge production are clearly the leading universities in the world, which are concentrated in the US and part of the same U-I coalition that includes big pharma. Again, patenting figures across the university show that the biosciences have not been the only departments to experience a marked increase in their connections with commerce (e.g. Mowery *et al.* 2004). But we have also seen that they do stand alone in the intensity of this development and this due to the exceptional dependence of biotechnology both on patenting and on 'basic science' research.

Each of the three dimensions (globalisation, commodification of knowledge and pioneer technology) thus shows a unique, but entirely contingent, compatibility with the demands of the three parties to the U-I biotechnology complex. Furthermore, we have seen above how each of these parties' success took off in 1980, the year of the structural shift of the monetarist revolution. Similarly, as regards the importance of the context of financialisation and its interaction with strong global IMPs, observers have also pointed to the central role of finance in the trajectories of the various parties. The importance of the demands from Wall Street for high profit margins in order to maintain share prices has been singled out as a major factor in the futile production of patented and branded me-too drugs by big pharma (NIHCM 2002: 4, 15). Kenney (1986: 133, 1998: 135) is also unequivocal that without a large venture capital (VC) sector, developed in particular on the back of the growth of computers in the 1970s, biotechnology would have struggled to launch itself.

Furthermore, Orsi and Coriat (2006) illustrate how (de)regulation of the New York NASDAQ stock exchange for hi-tech companies to allow the market listing of companies with no present profits but the promise of future profitability based on intellectual property assets (the 'Alternative 2' regulation) was an essential element of investment in the 'new economy' and a key link between strong global IMPs, especially for these hi-tech industries, and financialisation. Similarly, Zeller (2008) documents how the strengthening of IMPs created the regulatory conditions for the emergence of a positive feedback loop that has facilitated the massive accumulation of financial capital in the form of intellectual property rents (or super-profits) that are largely isolated from the vagaries and over-accumulation crisis of the productive sphere and hence, in turn, are especially attractive financial investments. In short, therefore, it was because of the total support of the historic bloc of finance capital that the patent coalition was so successful, but this level of support in turn is explained only by the unique two-fold active fitting together of the patent coalition with each other and with the structural imperatives of the global economy.

With the coincidental support of these structural imperatives, each of these three agents found that the mutual complementarities of their economic interests in the formation of strong links with the other two and the political implementation of their united wishes for US patent reform were mutually reinforcing. For their initial interests succeeded in eliciting patent reform that thereby facilitated the deeper integration of their interests in the formation of the U-I complex, which in turn presented a stronger political front for exacting yet further patent reform in their favour.

4.9 Conclusion

As big pharma obtained the domestic patent reform it sought, it became an even stronger political player, forging a *new* political agent and thus tightening its grasp over the machinery of the state; a crucial development for its success regarding TRIPs, for it was the USTR and not the pharmaceutical industry that had to drive this through.[38] This process, however, shows that the political success of patent reform was dependent upon the political support of all three parties from the outset, when the individual cases were weakest. This is particularly the case given the nature of the argument for patent reform: the linear model. For successfully arguing this case would be significantly impeded were the presumed parties to the process to challenge it.

This, therefore, highlights the crucial political role of the US universities in this process, because the *first* legislative measure of this process was the Bayh-Dole Act, authorising, and indeed encouraging, the patenting of publicly funded research results in universities (Slaughter and Rhoades 2002). Had the universities' interests been such that they themselves opposed this move, the entire process of the political strengthening of big pharma would certainly have been very much more difficult. As such, even if they are utterly indifferent or even against TRIPs, top American university life science departments were extremely important political players in that agreement's implementation. Furthermore, as major players with extensive bioscience patent portfolios, these departments stand to benefit greatly from TRIPs.

This analysis also allows us to explain why this was a particularly American development, because the radicalisation of the assertion of power of global capital will naturally be most prominent and most effective in the national economy at the system's hegemonic centre.[39] This, in turn, allows the particular forces at play in that society to be the major beneficiaries of this structural change. In this case, therefore, it was the American social forces of its dominant pharmaceutical corporations and its pre-eminence in molecular biology research (itself a peculiarly American discipline, as Kay (1993) and Kohler (1991) have shown) that were called upon and, using a suitably American metaphor, gladly 'stepped up to the plate'.

How, then, did TRIPs happen? This highly controversial international agreement was enabled by structural changes in the global economy towards a drive for primitive accumulation of knowledge resources. Capitalising upon this shift, an American U-I biotech complex, which stood to gain the most from it, was formed by the unification of big pharma, biotech start-ups and life science university departments in a singularly powerful single-issue political coalition in favour of domestic patent reform. Through their initial political successes in the US, regarding domestic patent reform and the legislative encouragement of U-I links, this coalition grew in coherence and power until big pharma, sitting atop this political enablement, could command the international trade machinery of the state at the centre of the global economy to drive through such global reform, fashioned in its own interests: the *actual* TRIPs agreement, privileging a global pharmaceutical industry and enforcing the patentability of biological materials, not just some generalised global patent reform.

But this analysis does not just illuminate the nature and provenance of the TRIPs agreement. By showing its connection to the ongoing commercialisation of academic research, the latter is also seen to be an indissoluble element of a much bigger transformation to the structure of the global economy, namely the neoliberal global primitive accumulation of knowledge. This has been for the benefit overwhelmingly of the profitability of US-based TNCs, as the geographical spread of the growth in patent royalties since 1980 exemplifies (see Figure 4.4; cf. Pagano and Rossi 2009).

As such, and against the claims of its protagonists, it is clear that TRIPs has *almost nothing to do with innovation* (which is itself usually uncritically valued positively) and its implementation simply cannot be understood if it is treated as such. Indeed, if anything it is increasingly apparent that strong patents *undermine* the development of innovation capacity not just in most developing countries but even in the US itself, leading to numerous calls from innovation economists for rebalancing of the patent system (Bessen and Meurer 2008, Boldrin and Levine 2010, Jaffe and Lerner 2007).

Moreover, the recent history of the political efforts to create a globalised, neoliberal knowledge-based (bio-)economy does not conclude, unfortunately, with the signing of TRIPs in 1994. This has been something of a surprise to many of the developing countries who finally submitted to signing TRIPs, for while they viewed it as the uppermost limit of their efforts to harmonise global IMPs, those behind the agreement saw it instead as merely the first step. TRIPs sets only minimum IP standards, leaves open to the discretion of national governments the actual form of many intellectual property laws and includes a number of provisions that provide limited flexibility for developing countries. For instance, compulsory licensing of drugs for national health emergencies is permitted.

Almost as soon as TRIPs was signed, therefore, moves were being made to start a new round of negotiations both to strengthen IMPs and to remove the flexibilities

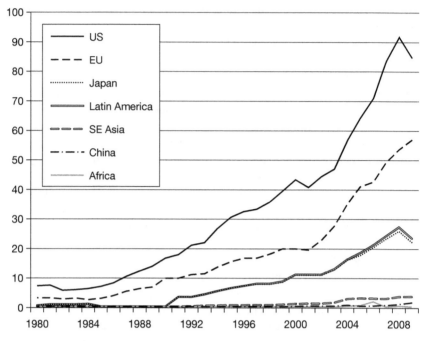

FIGURE 4.4 Receipts of royalties and licence fees ($ billions)

Source: Data from World Bank (2010), compiled by author.

of the TRIPs agreement; so-called 'TRIPs-plus' provisions (Sell 2007). However, growing outrage across the world about TRIPs' effects on access to essential drugs in developing countries had altered the political debate. The first significant altera-tion to TRIPs was thus the 2001 Doha Declaration, which confirmed that TRIPs could not be used to prevent the tackling of public health emergencies, i.e. in the *opposite* direction to TRIPs-plus.

Instead, therefore, TRIPs-plus efforts have been pursued not through the WTO but the WIPO (World Intellectual Property Organization) and bilateral trade agreements, mostly between the US or the EU and developing countries. As regards the latter, TRIPs-plus (and even US-plus) intellectual property provisions have been imposed upon various developing countries (especially in the Middle East and Latin America) through trade agreements, especially Free Trade Agreements and Bilateral Investment Treaties. The goal for such negotiations is explicitly the global harmonisation of IP law at the highest possible standard.

Regarding WIPO, this global harmonisation is pursued through the provision of 'objective' technical assistance to developing countries trying to establish effective intellectual property regulators (Drahos 2007, 2010), reflecting the fact that it is one thing to sign up to an agreement, another thing entirely to implement it (Deere 2009). The education, technical assistance and even socialisation of these fledgling bureaucracies that is provided by Geneva-based WIPO necessarily instils a model of intellectual property that both highlights the supposed benefits of IP, to the complete neglect of its downsides, and does so in a way that reflects the benefits of IP to a developed economy (Drahos 2007). Furthermore, given the vastly superior information handling capacity of global North patent agencies, most developing countries simply copy their decisions, as they have come to trust them during their training.

The drive towards the global privatisation of knowledge production through globally harmonised IMPs has thus continued apace to date, reflecting the continued neoliberal dominance of the global political economy. But just as the coup of TRIPs was dependent upon the rise of neoliberalism and the backing of bio-tech by finance capital, one may also expect that the current crisis of finance and neoliberalism will also have significant repercussions for the future trajectory of international IMP negotiations and biotech, and vice versa, themes we will explore in more detail in the closing chapter. Indeed, on some accounts, the strength of global and US domestic IPRs has been a significant, and entirely overlooked, factor in the very *generation* of the current economic crisis, due to the 'overpropertisation' of knowledge-intensive sectors together with the progressive 'upstreaming' of IMP claims towards ever-more abstract and 'basic' research findings (Pagano and Rossi 2009).

According to this argument, while the strong global IMPs of the TRIPs and TRIPs-plus agreements initially worked to the benefit of the national economy with the strongest national innovation system for science-intensive innovation (i.e. the US), in recent years it has also begun to come up against the limitations

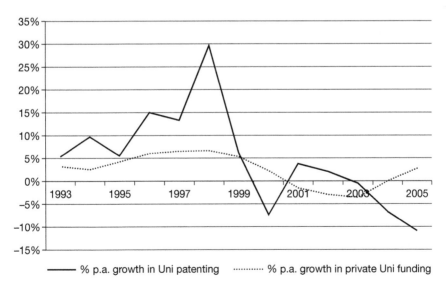

FIGURE 4.5 Annual growth rates for patenting and private funding at US universities (%)

Source: Data from USPTO (2010), compiled by author.

of this essentially parasitic model of capital accumulation, so that 'the patent system seems increasingly to be a source of uncertainty and costs, rather than a mechanism for managing and minimizing conflict' (Jaffe and Lerner 2007: 76). This has resulted in increasing paralysis even in the core of this financialised model of innovation, with *falling* rates of university patenting and private investment in US universities throughout the post-dot-com bubble period of the last decade (see Figure 4.5) and an increasing 'investment strike', even in IP-intensive industries (Pagano and Rossi 2009).

It is possible, therefore, that structural contradictions are now emerging with increasingly violent intensity, leading to ever more traumatic crises in the globalising knowledge-based economy. The future of this political economic project, including strong global IMPs, however, is clearly open and undecided though the trajectory of this unfolding history will continue to depend upon further constructions by and of structured agency. In this respect, however, one (set of) factor(s) is likely to be particularly significant, namely the rise of China. It is accordingly to this issue – and the commercialisation of Chinese science – that we now turn.

Further reading

Deere, C. (2009) *The Implementation Game: The TRIPs Agreement and the Global Politics of Intellectual Property Reform in Developing Countries*, Oxford: Oxford University Press.

Drahos, P. (2010) *The Global Governance of Knowledge: Patent Offices and Their Clients*, Cambridge: Cambridge University Press.

Drahos, P. and J. Braithwaite (2002) *Information Feudalism: Who Owns the Knowledge Economy?*, London: Earthscan.

Sell, S. (2003) *Private Power, Public Law: The Globalization of Intellectual Property Rights*, Cambridge: Cambridge University Press.

5

PRIVATISING CHINESE SCIENCE

National development versus neoliberal financialisation

5.1 Introduction

In previous chapters, we have constructed an explanation of the neoliberal commercialisation of science that has dominated the last few decades. But where are things headed from here? One possible glimpse of the future is a gleaming new facility, housed in an otherwise unprepossessing industrial tower block in Hong Kong (*The Economist* 2010c). This is the new laboratory of BGI (formally the Beijing Genomics Institute), which includes 120 state-of-the-art genome sequencing machines, or more sequencing capability than the entire US.

Most literature on the current commercialisation of science and the economics of science more generally focuses on the US. While this is justified to some extent by the process having proceeded first and farthest in the US, it also excludes consideration of processes of the commercialisation of science in other countries, including China. Moreover, the US-centricity distorts the debate, polarising between 'Mertonian Tories' versus 'Economic Whigs' (Mirowski and Van Horn 2005). The latter cheer the marketisation of science, seeing only a more efficient distribution of funding, while the former lament the passing of a 'golden age' of generous state support ensuring an open public sphere. Both sides of this 'state versus market' framing, however, discourage examination of the broader social forces that underpin science's commercialisation, taking the US political economic context as given and thereby limiting both the relevance of the resulting analysis in countries with completely different systems of research and innovation (R&I) and the future trajectories conceivable to social analysis.

Conversely, following the discussion of earlier chapters, we argue here that the commercialisation of science is indivisibly tied to profound 'structural' changes of financialisation, globalisation and the political project of neoliberalism. Furthermore, the future of these social forces is not fixed *ex ante* but irreducibly open, dependent upon ongoing political agency and concrete developments. This opens up questions regarding the mutual interaction of several parallel processes: the development of particular (techno-)scientific (inter-)disciplines; the commercialisation of that science; and the broader neoliberal project.

Again, the focus on the US may also be justified in terms of its seemingly unchallengeable dominance in both science and geopolitics. In both respects, however, the impact of China is likely to be a crucial factor conditioning the future 'economics of science'. For not only is China a rising global economic power, which will impact considerably on neoliberalism, but it is also rapidly improving its global standing in science. Furthermore, Chinese scientists are under significant and increasing pressures to commercialise research. The commercialisation of science in China appears similar to the much-discussed American process, regarding stronger links between enterprise and universities/public research institutions (PRIs, together URIs) and booms in the latter's patenting. They are also of a similar vintage: in the US, related to major structural changes around 1979/1980; in China, inseparable from the 'reform and opening up' of Deng Xiaoping's 1978 post-Mao reforms. The Chinese process has also been associated with familiar concentration on the life sciences, including agri-biotechnology.

These similarities, however, are belied by profound differences that challenge conventional understanding. Whereas the American experience has been at the very heart of the neoliberal project, neoliberalism has been a relatively superficial, contingent element of the Chinese state's drive to commercialise its science base. Its priority, rather, has been a project of national economic development towards China's reinstatement as a global economic power, i.e. 'catch-up'. 'Catch-up' has been approached pragmatically not ideologically. It has, in effect, involved the progressive transition from a socialist centrally planned economy to a capitalist market economy, though always subject to the imperative of preserving the political monopoly of the Communist Party (CCP). The specific reforms deployed are thus merely means reflecting the rolling redefinition of the goal of 'national development' in the light of contemporary political debate.[1]

Yet, by the late 1990s in the context of increasing exposure to the pressures of neoliberal globalisation, Chinese policy understanding of this goal had begun to solidify around the particularly neoliberal vision of a globally-competitive knowledge-based (bio-)economy (KBBE). It is in its commitment to 'catch-up' *thus defined* that China has become increasingly – though contingently – committed to neoliberal reforms, including privatisation of science. This gradual neoliberalisation, however, has remained secondary to the commitment to national economic development under the CCP's leadership and has also always been mediated by the practical interpretation and implementation of policy, often by local governments (Segal 2003); i.e. a qualified embrace of neoliberal policy 'to *sustain* a restructured, recalibrated developmental state that could respond to the perceived imperatives of the globalizing, knowledge-based economy' (Jessop 2001: 43, emphasis added) rather than abandon it altogether.

In short, we argue that (1) science policy in China has become progressively more neoliberal in appearance but (2) the resulting commercialisation of science remains markedly different to the quintessentially neoliberal case of the US. These differences have important, if as yet unclear, implications for the co-evolution of science and neoliberalism. Moreover, understanding these differences has crucial theoretical and methodological lessons for an economics of science (and science and technology studies (STS)) in that a political economy perspective is needed to make sense of these crucial differences.

5.2 The uniqueness of the US and the challenge of China

Let us start by explaining more fully the choice of the two countries compared in this analysis. American science has been progressively privatised in interactive parallel with a particular conjunction of political economic changes since the 1980s. These have been most marked in the US, but given its central position in the global capitalist political economy, they have had global impact. As discussed in earlier chapters, built on the ongoing over-accumulation crisis since the late 1960s, finance capital launched a political 'counter-revolution' in 1979/1980, seizing power over the US state (and thence the global economy) and transforming the priorities of governance from Keynesian welfare/warfare regulation to financial profits (Arrighi 1994, 2008). This allowed finance, as a Gramscian historic bloc, to compel the restructuring of the global economic architecture to enable the appropriation of new resources capable of delivering super-profits. Knowledge production and – facilitated by new scientific techniques – biological resources have been the primary targets of this primitive accumulation or 'accumulation by dispossession' (Harvey 2003, Slaughter and Rhoades 2004, Zeller 2008). The result, in the following three decades, has been the 'financialisation' of the economy – i.e. the economic and *political* dominance of finance over the political economy (Arrighi 1994, Blackburn 2006) – and the enclosure of knowledge and biological commons.

These changes presupposed both committed agency and favourable structural conditions, both of which were most propitious in the US. The political agency may be summarised under the term 'neoliberalism'. Rising to political prominence with the crisis of the Keynesian spatio-temporal fix of capitalism (Jessop 2002a), it has been the ideological driver for, and political movement behind, the subsequent 'unleashing' of capital (Glyn 2006).

Furthermore, privatising science is a *central* element of neoliberalism, based on the idea of the 'marketplace of ideas' (Mirowski and Sent 2002a). The market fundamentalism of neoliberalism arises from the Hayekian argument that a market is primarily an epistemic, and not an allocative economic, social mechanism. The wonder of the market, according to this argument, is that it produces the socially optimal output (or 'conclusion') from individual inputs that necessarily have incomplete information. No social planner or government agency, therefore, can *possibly* have as much information as a market outcome, while the latter also maximises individual negative freedom. But as an epistemic mechanism, it follows that science – the social epistemic practice par excellence – should also be structured as a free marketplace of ideas, complete with price signals and open monetary competition. Neoliberalism and privatising science are thus indivisibly connected.

Neoliberalism, however, is a contradictory assemblage – not a unified, monolithic and coherent social force. Nor does it have *sui generis* causal powers in isolation from concrete agency, but rather assumes different manifestations in particular sociohistorical contexts and in their differing responses to the complexities and contestations intrinsic to neoliberalism (e.g. Nonini 2008, Peck and Tickell 2002). It follows that the US experience of neoliberalism and privatising science cannot legitimately be treated as a universal phenomenon. In particular, the privatisation of American science is *unique* as a central element of the neoliberal project of US-led, financialised primitive accumulation of global knowledge production, exemplified by the World Trade Organisation (WTO) Trade Related Intellectual Property Agreement (TRIPS) on intellectual property (May and Sell 2006, Sell 2003).

Since the privatisation of American science must be seen as inseparable from changes in the global political economy, however, a particularly striking development in the latter must be confronted, namely the rise of China. Not only is the growth and global integration of China's economy arguably an epochal shift, but Chinese science is also rapidly growing in global significance (Jakobson 2007, Wilsdon and Keeley 2007, Zhou and Leydesdorff 2006) (see Table 5.1).

Certainly, many of these figures demand careful analysis, and not just because of unreliable statistics. Such is the size of China and the low level from which it has come, especially following the devastation of the Cultural Revolution, that Chinese science data are a 'hall of mirrors' (Leadbeater and Wilsdon 2007). Stunning absolute figures may be meagre in per capita terms while middling absolute performance masks extraordinary rates of growth. Indeed, it is particularly difficult to assess the overall progress of Chinese science (Evidence 2007).[2] Some assessments see China as doing well, even as an ascendant science 'superpower' (Sigurdson 2005), while others argue that it still has major weaknesses (Fischer and von Zedtwitz 2004). Nevertheless, there is little disagreement that China is becoming increasingly important in both the global political economy and science. In both respects, the case of China has important lessons for the future co-evolution of science and neoliberalism, and so for the broader process of privatising science.

Indeed, China is singularly important for a contemporary economics of science for at least four further reasons. First, there is the recent but growing trend for Chinese-based institutions to be incorporated into the 'global innovation networks' (GINs) of major transnational corporations (TNC) (Ernst 2008). This would also include the increasing number of TNC research and development (R&D) centres in China; facilities, moreover, that are not just conducting minor product adaptations or even shells established as a necessary quid pro quo for market access at the insistence of government (*Economist* 2010b). To be sure, China's science and innovation capabilities still hugely lag those of the rich global North, yet there is clear, if contradictory, evidence of its significant improvement (Altenburg *et al.*

TABLE 5.1 China's recent S&T indicators

	1997	2001	2005
General expenditure on R&D (GERD) (US$ billion)	6.1	12.5	30.1
GERD/GDP (%)	0.64	0.95	1.34
Government S&T appropriation (US$ billion)	4.9	8.4	16.4
Government S&T appropriation/ total government expenditure (%)	4.4	3.7	3.9
Scientists and engineers (1000 full-time employees)	588.7	742.7	1,119
R&D personnel (1,000 FTEs)	443	957	1,365
S&T personnel (1,000 FTEs)	1,474	3,141	3,815
Graduates in science, engineering, agriculture and medicine from HEI undergraduate courses (1,000)	496	556	1,528
S&T papers catalogued by SCI, ISTP and EI	35,311	64,526	153,374
Share of total global OSI publications (%)	2.5	3.7	6.0
Share of world citations (%)	0.92 (1995)		3.78 (2004)
Domestic patents granted	742	2,468	14,761
Exports of hi-tech products (US$ billion)	16.3	46.5	218.3
Hi-tech products/total exports (%)	8.9	17.5	28.6

Sources: Evidence (2007), Guan *et al.* (2005), Jakobson (2007), MOST (2007), Sigurdson (2005) and Wilsdon and Keeley (2007).

2008), including in key science-intensive sectors such as genomics, nanotech and space science.

Second, the flipside to the super-profits of innovation rents that have accrued predominantly to US-based TNCs and their financial investors is the seemingly limitless cheap and comparatively educated Chinese workforce. This has produced an international division of labour, of both production and innovation, in which Chinese firms have been largely excluded from these gains and instead are increasingly exposed to cut-throat competition over razor-thin profit margins that, in turn, preclude investment in innovation upgrade (Steinfeld 2004). As a result, the national government's utter policy priority of 'climbing the value chain' is in direct tension with continuation of the current neoliberal structures of global economic governance.

The growing geopolitical power of China thus raises serious challenges to the continuation of the neoliberal regulatory architecture of the global economy (most notably TRIPs), which is so heavily weighted against China's techno-economic 'catch-up' (e.g. compare the receipts of royalties and licence fees by the US and China in Figure 4.4 on p. 114). This is further exacerbated by the final two points, namely: the heated debate regarding technology transfer of proprietary low-carbon and environmental technologies owned by companies in the global North to assist with the globally urgent task of expediting China's low-carbon transition; and the fact that the rise of China represents the growing global influence of an economic model involving strong, even authoritarian, state leadership that also directly challenges neoliberal policy prescriptions.

Furthermore, the rise of China is particularly significant for science (and vice versa) given the singular importance the Chinese state has accorded to the issue ever since the imperial reformers of the mid-nineteenth century (Elman 2007: 523). In the current period, Deng Xiaoping initiated the process of 'reform and opening up' by listing 'science and technology' (S&T) as one of the 'four modernisations' (with agriculture, industry and defence). Modernisation of the science and innovation system has remained at the heart of government policy reforms (Segal 2003), seen as of equivalent importance to the Special Economic Zones and the construction of Shanghai's Pudong international business district (Cao 2004). This has been manifest most recently in the political slogans of the 2006 Medium to Long-Term Plan of 'scientific development' (*kexue fazhan*) and 'indigenous innovation' (*zizhu chuangxin*) (Schwaag-Serger and Breidne 2007, Xue and Forbes 2006).

Despite the importance of science to China's modern history, however, studies of Chinese science, including in Chinese, concentrate on its ancient achievements and effectively neglect the politically sensitive issues surrounding post-imperial science (Fan 2007). There is thus a real need for concerted reflection on modern Chinese science in STS. Furthermore, analysis of China offers particular insights regarding the privatization of science because of the challenge it poses to the US-centric narrative; one that, moreover, is itself implicated in the neoliberal project. Understanding this process, however, depends upon situating it within the broader process of political economic reform in China, which in turn shows how the global neoliberal pressures have been filtered and transformed by the specific Chinese context.

In particular, four major themes problematise the presumptive neoliberalism of ostensibly familiar trends regarding both the broader economic reform process and commercialisation of science. We explore each of these themes in the evidence below. They are:

1   The dominance of a national project of economic development, under the unchallenged leadership of the CCP.
2   Political leadership of these economic changes by the central government itself, not by a capitalist historic bloc.
3   The blurring of state and market, with the former straying into the latter, not vice versa (as in the West) – what Duckett (2001) has called 'state entrepreneurialism'.
4   A progressively embedded commitment to neoliberal policy prescriptions, including the whole project of a 'knowledge-based (bio-)economy' (KB(B)E), though in a way that remains fundamentally conditioned by issues 1 and 2 and undermined in practice, especially by local implementation, given China's relatively devolved political structure (see Table 5.2).[3]

TABLE 5.2 Phases of China's economic reforms and of neoliberalism

Dates	China's economic reform		Phases of neoliberalisation
	Reforms	*Characteristics*	
Pre-1978	**Pre-reform Maoism**		**Proto-neoliberalism** Theoretical critique of Keynesianism
1978–85	**Early experimental reforms** Rural economic reform via Household responsibility system and Township–Village Enterprises (TVEs)	**'Growing out of the plan'** Zhao Ziyang's leadership: cautious, consensual decision-making Introduce markets where feasible: focus on agriculture and industry	**Roll-back neoliberalism** Deregulation and structural adjustment
1985–92	**Non-capitalist market economy** Decollectivisation of farms completed, urban economic reform begins, dual-track pricing, gradual replacement of planning by market	No privatisation Decentralise authority and resources 'Reform without losers'	
1992–97	**Post-'Southern Tour' reforms** Promotion of Special Economic Zones, shrinking of state sector and employment, restrictions on size of private enterprise and foreign investment lifted	**Concerted marketisation** Zhu Rongji's leadership: rapid, personalised decision-making Strengthen institutions of market economy, focus on finance and regulation	**Roll-out neoliberalism** Reregulation in favour of capital
1997–2009?	**Post-'Asian Crisis' reforms** WTO membership agreed and implemented, FDI mushrooms, property rights clarified, 'National Champions' expanded to include private business	Beginnings of privatisation Recentralise resources, macroeconomic control Reform with losers	

Source: Naughton (2007: Table 4.1) with Peck and Tickell (2002).

In short, Chinese economic reform, including the commercialisation of its research base, has been part of an ongoing attempt to develop a globally competitive national economy *in conditions of neoliberal globalisation* that make it 'even less feasible than in the past for any single country to pursue an idiosyncratic technological path . . . isolated from external S&T trends' (Gabriele 2002: 336). The Chinese government has thus increasingly identified the national project of development with the KBBE and shifted towards commensurately neoliberal policies, including the privatisation of science. It is the national, top-down project, however, that remains paramount, and the privatisation of Chinese science has thus taken a very different route to parallel developments in the US.

To be sure, both the conditioning of the actual implementation of neoliberal prescriptions by domestic political structures and the strong role of the state in economic development are not uniquely Chinese phenomena, but rather common features of most (if not all) East Asian economies, notwithstanding the important differences among them. We focus our comparison on China and its particularities here, however, not just to analyse a single case study in sufficient detail (recalling the 'double challenge' discussed in Chapter 1), but also because of its presumptive importance as rising global power, including in science, that poses particular and current challenges to the US-centric neoliberal order.

5.3 Privatising Chinese science

In recent years, Chinese science has been greatly exposed to commerce. This process shares many features with those in the global North, but many of these similarities are superficial. To understand this contrast, we must first consider the starting point of China's reforms (see Table 5.3). From the 1950s, China's science system was modelled upon that of the USSR. Science was organised vertically into disciplinary fields under the leadership of a ministry, which also controlled the relevant state-owned enterprises (SOEs). Universities and public research institutes (PRIs) (the latter doing the best 'basic' research) utterly dominated the 'national innovation system', with SOEs doing little, if any, R&D. Furthermore, the system offered few incentives for collaboration – and many reasons actively to avoid it – and little or no incentive to innovate (Liu and White 2001). The challenge for Chinese reforms regarding science and commerce, therefore, has been almost the exact reverse of that in the global North (OECD 2008: 64): trying to get *commerce into science* rather than science into commerce, i.e. trying to get business to do science not trying to expose science to commercial pressures.

The four non-neoliberal aspects of the reform process are clearly in evidence. First, the overarching goal of the reforms has been the national project of economic development, with commercialising science simply a means, albeit an integral one, for economic policy (Kroll and Liefner 2008). The state has taken the lead through a series of characteristically Chinese 'plans' (see Table 5.3, in bold) (Sigurdson 2005), adopting a clear 'picking the winner approach' (OECD 2008: 84), the very antithesis of neoliberal science and technology policy prescriptions. These plans have concentrated finances on improvement of leading national research centres, while offering considerable managerial and financial autonomy to funding recipients (Gabriele 2002); a mixed and flexible approach aiming to 'walk on two legs'. It has also provided platforms to attract non-state funding. For instance, only 2–5 per cent of funding for the Torch high-technology plan has come from the government (OECD 2008: 84).

Considerable state/market blurring is also apparent, with the state involved directly in enterprise and the market not necessarily 'private'. It is thus especially important to distinguish between the *privatisation* of science and its *commercialisation*; these terms cannot be considered interchangeable in the Chinese context. Finally, regarding science reform one can also clearly see a progressive shift towards neoliberal KBBE policy, i.e. the shift from commercialisation to privatisation. In recent years especially, China's science policy rhetoric has converged with that of developed countries as government scientists have imported the discourses of science parks and clusters (Zhu and Tann 2005); what Sum (2004) calls the 'siliconization' of China as government attempts to create its own Silicon Valley. This yields a clear periodisation of science reform that maps closely onto that of economic reform more generally (e.g. Kroll and Liefner 2008; Table 5.3).

Let us now examine particular areas of this reform process to show (1) the apparent progressive neoliberalisation of policy in *form* (shifting from commercialisation to privatisation) and (2) its marked differences in *substance* to more familiar processes of privatising science. Science-industry links may take three forms paradigmatically, namely spin-offs, commercial funding and patent licences, in ascending order of arm's-length distance between parties. We examine each of these in turn.

1. Spin-offs

Progressive neoliberalisation

Spin-offs directly from universities and research institutions (URIs) alone were feasible in the early reform period, while the other two forms only became significant from 1998/1999 (Kroll and Liefner 2008). The form and importance of spin-offs have also changed in recent years, shifting from commercialisation to privatisation of science.

The science reforms began with a system dominated by URIs and structural disincentives to collaborate. Regarding spin-offs, the first stage of reform (1985–98) was characterised by the emergence and flourishing of a unique Chinese phenomenon: the university-owned technology enterprise (UOTE) (Chen and Kenney 2007, Wu 2007). Chinese universities have owned enterprises since the 1950s, though often only service companies or student employers (Wu 2007). After budget cuts of up to 70 per cent in the mid-1980s (Segal 2003) (see below), however, URIs began seeking extra funding by increasingly setting up UOTEs. Revenues grew considerably from the late 1980s (RMB 1.76 billion in 1991 to RMB 37.9 billion in 1999 (OECD 2008)) and continued into the new millennium;

TABLE 5.3 Timeline of China's science reform

Early reform period	
1978	Eight-Year Plan for Science and Technology
1980	Chinese Patent Office founded
1982	**'Key Technologies R&D Programme'** to develop technology urgently needed for industrial upgrading

Initial commercialisation reform	
1985	Decision on Structural Reform of the Science and Technology System begins the reform process
	Decision on Reform of Educational System
	China's Patent Law enacted
1986	**'863' Hi-tech Research and Development Programme** initiated (launched March 1987) to enhance China's international competitiveness in six hi-tech fields, e.g. IT, biotech and new materials
	'Spark Programme' approved to promote rural economic development based on S&T
1987	China's Technology Contract Law introduced to facilitate technology transfer
1988	**Hi-tech 'Torch Programme'** launched – to develop new technology industries and hi-tech development zones
	Forerunner of Zhongguancun founded as science park
1989	Six documents of official ministerial guidance issued to deal with issues related to technology markets (1989–91)
1990	Copyright Law introduced
1991	Ten Year Programme of S&T Development 1991–2000 launched
	Central government officially approves university-owned enterprises/start-ups
	First 'Science & Technology Industry Park' established

Concerted marketisation reform	
1992	**National Basic Research Priorities 'Climbing Programme'** introduced to strengthen China's basic research capabilities in selected areas as a solid foundation to tackle significant socioeconomic problems
	S&T Progress Law introduced structural adjustments towards a socialist market system
	State Basic Policy for Hi-Tech Industrial Development Zones
	Patent law substantially revised expanding protection
	Government-financed venture capital firms established in four cities (1991–93)
1993	'Decision on Several Problems Facing the Enthusiastic Promotion of Nongovernmental Technology Enterprises' issued, offering official recognition of this crucial business form
	'Law on the Progress of Science and Technology' introduced to encourage R&D international collaboration
	'211' Programme launched to create 100 world-class universities for the twenty-first century, oriented to economic goals and introducing competition into HE sector
1994	National forum triggers wave of mergers and institutional reform in HE
1995	S&T reform process redoubled with 'Decision to Accelerate the Development of science and Technology' and 'Decision on Profound S&T System Reform'
	'Suggestions on Deepening Higher Education Structural Reform' policy document released
	Education Law confirms the state's full support for establishment of private schools and HEIs
1997	**'973' National Basic R&D Programme** starts, absorbing the 'Climbing Programme'

TABLE 5.3 continued

Privatisation unleashed

1998	Individual researchers in government-funded R&D projects allowed to obtain a royalty of up to 35% of the licensing fees
	Announcement No. 1 at the Ninth Conference of the National People's Congress allows establishment of corporate and foreign VC firms
	Chinese Academy of Sciences 'Knowledge Innovation Programme' to create 30 internationally recognised PRIs by 2010 initiated. Restructuring of PRI sector begins
	'100 Talents' Returnee Scheme initiated, to attract Chinese scientists based overseas to return
	'985' Programme initiated to create elite group of globally competitive HEIs
	National Higher Education Law sets the parameters of universities' autonomy
1999	National Innovation Congress organised by central government (at which . . .)
	State Council approves 'Several Provisions on Promoting the Transformation of Scientific and Technological Achievements' giving definitive support to the entrance of private firms to high-technology fields. 'In contrast to the vagueness in the 1993 and 1995 State Council decisions, the 1999 decision called for concrete measures to foster high-tech industries and included a fund to support S&T innovations by small- and medium-sized enterprises and preferences for domestic high-tech products and equipment in government and enterprise procurements.' (Segal 2003: 38)
	Chinese version of Bayh-Dole Act issued by Ministry of Education
	Beijing's Zhongguancun science park approved by State Council and cited as of equal significance to Shenzhen and Pudong
2000	University-backed VC firms begin to emerge
	Patent law amended for WTO compliance, simplifying application, grant and rights transfer and abolishing special treatment for SOEs
2001	National Technology Transfer Centres (NTTCs) established by Ministry of Education (MOE) & State Economic and Trade Commission (SETC) to promote the commercialisation of technological achievements at 6 leading universities: Tsinghua University, Shanghai Jiaotong University, China East Polytechnic University, Huazhong S&T University, Xi'an Jiaotong University and Sichuan University
	The State Economic and Trade Commission upgraded to ministerial level as the Ministry of Science and Technology (MOST)
	MOST and MOE certify 22 national-level university science parks
	First provisional regulation ('Alignment of the legal and financial systems with a market-oriented business approach') issued to foster an institutional environment favourable to investment in new enterprises
	New law issued granting 'unprecedented flexibility of Beijing's rigid household registration system so that companies in ZGC can now hire college graduates regardless of their original household registration. This is a landmark change for a city that has practiced some of the world's strictest migration controls for forty years.' (Zhou 2005)
2002	MOE issues clear directive encouraging the development of university enterprises
	MOST and MOE certify further 21 national-level university science parks

'Indigenous innovation'

2006	**Long-to-Medium Term Plan** announced, prioritising 'scientific development' and 'indigenous innovation', with GERD to reach 2.5% of GDP by 2020 (approximately US$113 billion) (Wilsdon and Keeley, 2007)
	11th five-year plan also prioritises science and technology

the top ten universities' UOTE revenues grew from RMB 18.1 billion in 1999 to RMB 475.8 billion by 2003 (Chen and Kenney 2007: Table 3).

UOTEs have been crucial for commercialising Chinese science, and a relatively successful policy vis-à-vis the goal of national development but they are in marked contrast to the neoliberal privatisation of science. As these URIs are themselves effectively 'state' bodies, the UOTE is an example of the state/market hybrid of 'state entrepreneurialism' (Duckett 2001). The exact status of property rights has also remained deliberately and systematically ambiguous, clarification and express 'privatisation' only occurring in recent years.[4]

This second stage of spin-off reform, from the late 1990s, has been marked by the increasing *privatisation* of science. Private ownership of spin-offs has been allowed since 2000, leading to a 'wave of private spin-off formation' (Kroll and Liefner 2008: 305). Conversely, UOTEs have become less important. Numbers grew dramatically up until 1995 but then contracted, especially after 2000, even as revenue growth continued (Kroll and Liefner 2008, OECD 2008).

The rise of private spin-offs has also involved increasing attention to familiar Western policy recommendations, in particular regarding science parks (Prevezer 2008). Once again, the late-1990s watershed is crucial, the government only becoming 'properly involved' in fostering science parks around 1998/1999 (Cao 2004) when Beijing's flagship Zhongguancun science park was officially established. Science parks then mushroomed, 43 of the 49 national-level university science parks being authorised in 2001/2002 (Hu and Matthews 2008, OECD 2008)

Differences

Privatising science this way, however, remains difficult and relatively superficial. While private spin-offs are growing, interest in the science base from domestic Chinese industry – potential future purchasers of spin-offs – remains lacking. And while science parks have appeared near every research university, many are simply lucrative real-estate deals called 'science parks' to please bureaucratic bosses (e.g. Wu 2007).

Most importantly, the fundamental institution of venture capital (VC) remains underdeveloped. Chen and Shih (2005) conclude uncompromisingly that 'China still has no effective mechanism for high-risk investment' including at 'most high technology development zones'. Such VC as there is has been dominated until very recently by government- and university-run funds (Cao 2004, OECD 2008); a set-up that is usually inefficient and inexpert and a further example of state/market blurring. Indeed, UOTEs initially fulfilled the role of VC 'in an environment that in the early 1980s could not have supported true VC either ideologically or economically' (Chen and Kenney 2007: 1065), forging links between research and commerce without mass lay-offs. Today, however, the lack of VC crucially constrains the progressive shift from UOTEs to private spin-offs.

2. Commercialisation of funding

Progressive neoliberalisation

The second form of science-industry connection is direct funding of research by businesses. Three developments are particularly significant, again concentrated in the late 1990s. First, from the outset of science reforms in the mid-1980s, Chinese URIs faced drastic cuts of government funding, at 5 per cent per year from 1986–93 (Hong 2007, quoting Zhou *et al.* 2003). This stimulated a search for other sources of funding, including the rush of UOTEs. However, unlike neoliberal policy aiming to stimulate private research enterprise, in China the shift to commercialisation was 'less a result of targeted research policy than a result of the state's general retreat from regulation and funding in the research sector' (Kroll and Liefner 2008: 300).

The second important development followed in the late 1990s, as the overall funding of science in China shifted to the business sector (see Figure 5.1). In fact, the percentage of total R&D funds invested in business fell in the early 1990s. In the aftermath the Asian financial crisis of 1997 and the backroom agreement on WTO membership (Naughton 2007), however, a series of policies were introduced to stimulate the commercialisation of science and the redoubled concentration of resources for elite research institutes (see Table 5.3, pp. 126–127).

These policies succeeded in causing a boom in business funding and a dramatic reversal in the balance of funding between URIs and business, from 53 per cent versus 45 per cent respectively in 1998 to 37 per cent versus 61 per cent in 2002

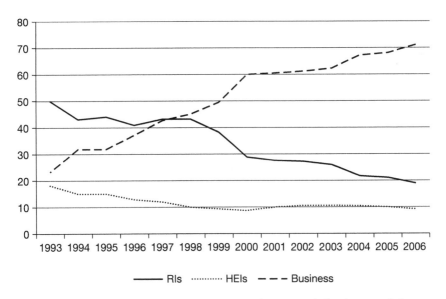

FIGURE 5.1 The transformation of Chinese R&D from URI-led to business-led (% total R&D appropriation by institutional type)

Source: MOST (2007).

(Guan *et al.* 2005). This trend has continued to date: business funding more than doubled between 2003 and 2006 from RMB 92.5 billion (60.1 per cent of total R&D funding) to RMB 207 billion (69.1 per cent) (Chen and Kenney 2007, MOST 2007).

Differences

However, the vast (and increasing) majority of this business funding has been directed to business itself. Similarly, while business funding now steadily constitutes about 36 per cent of university funding, it accounts for only 3.1 per cent of PRIs' budgets in 2006, down from 5.3 per cent in 2003 (MOST 2007). Conversely, government funding has grown from its lows in the 1990s, at around 20 per cent per annum for the past five years at the National Natural Sciences Foundation of China (NSFC) (personal communication).

Furthermore, the meteoric rise of the business sector in Chinese science is itself something of an artefact rather than a genuine social transformation, for the third significant development regarding funding reform is the restructuring and reclassification of the PRIs, again in the late 1990s. 'From 1998 to the end of 2003, 1,149 public research institutes were converted into business entities' transferring '111,000 S&T personnel . . . to the business sector' (OECD 2008: 200). As a result, the business sector received a sudden, but government-administered, growth-spurt. Yet, the S&T expenditures of China's large and medium-sized enterprises actually shrank during the same period (ibid.). Thus, while there has been a leap towards the privatisation of Chinese science in the last ten years, there is little evidence of take-off in reciprocal participation of private enterprise in science.

3. Patenting

Progressive neoliberalisation

The rise of patents is one of the most notable features of neoliberal science reforms. Patenting in search of private profit by URIs is ideotypically neoliberal, creating a market in research results (see Chapter 4). Furthermore, as the form most dependent on arm's-length market transactions, and hence presupposing relatively strong and reliably enforceable property rights and a private business sector interested in *licensing* the patents, URI patenting depends upon other crucial elements of a neoliberal science system.

China's relative lack of 'respect' for intellectual property, especially copyright, has engendered much consternation from Western business. The Chinese government, however, has made concerted efforts to tighten intellectual property law, especially recently to make it TRIPs-compliant (China having joined the WTO in 2002), and to improve enforcement. In parallel, patenting has become an increasingly important science-industry link but only in the latest phase of economic reform, i.e. following the 'privatisation unleashed' reforms of the late 1990s (see Table 5.3). Even as regards this quintessentially neoliberal reform, however, there are significant differences from familiar processes in Western countries.

First, consider the periodisation of the transformation of URI patenting in China. Three stages are apparent. Domestic Chinese patents were *initially* dominated by the URIs, reflecting their dominance of the research system from the pre-reform era; an additional example of state/market blurring and a striking difference to the US situation. The second stage was the transition from URI to business domination (1992–98). Patenting remained low throughout the 1990s (Kroll and Leifner 2008) and the percentage of total patents from URIs, having fallen consistently from 1985 to the mid-1990s, remained small but stable from 1993–2000 (Motohashi 2008: Figure 7). Only in the third stage (1998 onwards) has patenting become a significant form of science-industry link, in particular following the 2000 TRIPs revision of Chinese patent law and the 1999 enactment of legislation equivalent to the US Bayh-Dole Act, allowing URIs to patent publicly funded research.

From 2000 there has been a 'surge' of both Chinese patents by domestic entities and US patents by Chinese entities (Figure 5.2) (Chen and Kenney 2007, Guan *et al.* 2005, Hu and Matthews 2008). Business patents rose most quickly, increasing over 50-fold between 1998 and 2006, but those from both PRIs and universities have also increased, at over 7.5- and 25-fold respectively. As a result, the percentages of patent grants from business and PRIs have effectively swapped places, with universities at levels near their initial peak in the mid-1980s (Figure 5.3) (Motohashi 2008).

Invention patent applications and grants per 1,000 R&D personnel also demonstrate significant growth in patenting activity. These rose slowly between

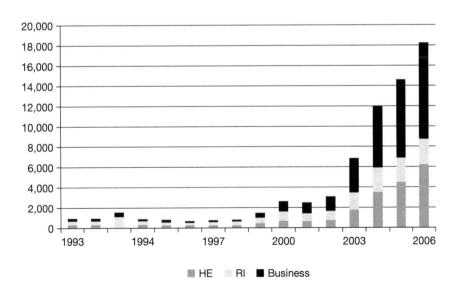

FIGURE 5.2 Domestic invention patents granted, 1991–2006

Source: Guan *et al.* (2005) and MOST (2007).

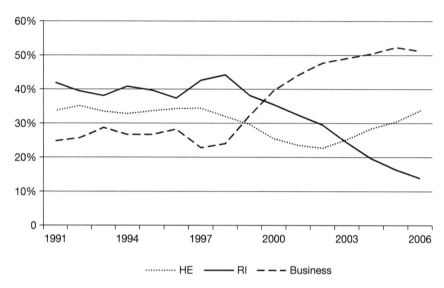

FIGURE 5.3 Share of invention patents granted by institutional type, 1991–2006

Source: Guan *et al.* (2005) and MOST (2007).

1994 and 1998 and then strongly from 1999; the former from 3.8 in 1994 to 31.3 in 2005, the latter from 1.4 to 11.3 (ibid.). Furthermore, although the percentage of invention patents granted to PRIs has fallen, both the absolute number of grants and the number of grants per R&D expenditure have increased, again especially since 2000. Thus, while PRIs have increasingly shifted towards 'basic research' following the 1998–2000 restructuring, they have also intensified their patenting activity.

The adoption of patenting as a performance indicator for national science and technology (S&T) programmes has been another key reform (OECD 2008). Similarly, domestic university rankings, upon which government funding is based, have included patents as an important criterion since around 2000. Leading universities and PRIs have given faculty training in IP management in order to instil a patenting culture. Technology transfer offices have also appeared in most URIs with a rush of 'offices for commercializing science and technology achievements' (STACOs) in the late 1990s and national technology transfer centres (NTTCs) in 2001 (OECD 2008: 193). Both STACOs and NTTCs are tasked with facilitating technology transfer, especially patenting, the difference being that the latter can also work on international technology.

Further indications of the transformation of patenting in China include the growth of foreign patentees since 2000, given the greater ease and reliability of the system after TRIPs reforms, and the increasing quality of domestic patents, shifting towards the more technologically sophisticated 'invention' category, particularly at URIs. Finally, over 90 per cent of lawsuits are now between domestic parties, an indication of the growing strength of China's patent system.

Differences

URI patenting, however, remains very different from familiar Western trends. The most obvious difference comes from an examination not of URI patenting but of its apparent complement, i.e. the continued relative weakness of business patents, despite their recent growth. Only in recent years have business patent grants exceeded 50 per cent of the total domestic patents. Patenting remains of marginal importance, at best, to the majority of Chinese businesses (Wilsdon and Keeley 2007). URIs also remain more adept patentees than business, hence their percentage of invention patent *grants* is sizeably larger than the percentage of *applications*, while the opposite is the case for business (Motohashi 2008).

Similarly, patent licences have yet to become a major mechanism of technology transfer from science to industry (Wu 2007). URI patenting grew significantly after 1999 but revenues from patent contracts grew much more slowly (Kroll and Liefner 2008). The patenting activity of URIs is also easily exaggerated. Increased patenting by URIs has resulted everywhere in a concentration of gains and China is no exception. While the leading universities had only 35 per cent of patents giving rise to sales, they recouped 84 per cent of the real licensing income (OECD 2008). Furthermore, the penetration of patenting practice into URIs remains rather shallow, with 'only about 10% of university research . . . put to any kind of commercial use' (Wu 2007: 1078).

These differences from neoliberal expectations may be readily explained by a combination of three interrelated factors, namely: the particular structural conditions of the Chinese economy; the default cultural treatment of intellectual property; and the determination of the government to maintain control over the development of the science system. The first relates to the absence of a strong private enterprise sector interested in, and so politically mobilising for, the licensing of patents from URIs. Not until there is a powerful, domestic constituency demanding strong patent rights will they be effective (May and Sell 2006). Meanwhile, the structure of the Chinese economy and the demands of the national 'catch-up' project may well demand a much looser IP regime to encourage learning from internationally-leading technology firms – as in successful historical precedent (Chang 2002).

Second, intellectual property was simply an alien concept at the outset of the reform period, with S&T developments considered a 'public good belonging to the whole nation' (Fischer and von Zedtwitz 2004). The vertical structure of accountability of SOEs and URIs to government embedded this view through conditioning of daily work practices. Introducing a patenting culture has thus been a long, difficult process involving profound institutional reform. This process remains incomplete and contested. For instance, while assessment of faculty increasingly measures patents, 'faculty promotion guidelines continue to give much less credit to commercialisation than to scholarly publications' (Wu 2007: 1088), and there are often financial incentives for the latter too.

Third, while leading the reforms, the central government has harboured deep ambivalences to strong intellectual property rights, for reasons just discussed. It is also determined to maintain control over the broader economic reform process. Not only is patent activity thus often compromised by government interference, but the ascendancy of a powerful, capitalist class made wealthy through patenting success is discouraged. Similarly, increased patenting in Chinese science has not inaugurated an open international market in patent licences but remains a difficult issue for international collaboration (Wang and Ma 2007). Foreign-funded R&D centres and joint projects have grown since the 2000 watershed, but joint patents and patent licences remain overwhelmingly domestic affairs. In 2005, of joint university-industry patents, 81.1 per cent had both Chinese partners, against 1.7 per cent between Chinese universities and overseas businesses and 0.1 per cent between Chinese businesses and overseas universities (Motohashi 2008). Personal interviews have also revealed several cases in which approaches by foreign businesses to URIs have been obstructed by local government despite faculty support.

In short, patenting in China has grown due to government policy commitment but merely as a *means* to catch-up, and thus always subject to possible trumping where patent/licence activity is perceived to conflict with that over-riding goal. Patents are also thus put to quite different use in China from the expectations of neoliberal theory, namely to defend national technological advances against global competition and not to stimulate and incentivise innovation by private entities (Hu and Matthews 2008: 1469).

4. Summary

In conclusion, the deepening of science-industry links has proceeded apace in China, with a progressive shift from policies encouraging the *commercialisation* of science to those encouraging actual *privatisation*. This is reflected in the shifting balance of importance of the three forms of science-industry link, with UOTEs the least dependent upon privatisation and patents the most. However, even to the extent that there has been a progressive neoliberalisation, reforms have remained beholden to, and qualified by, the overarching goal of national economic development under the CCP. Hence, each of the four themes 1 to 4 (see p. 122) are clearly in evidence.

5.4 China and the US compared

We finish with a comparison of the commercialisation of Chinese and American science.[5] The two processes share numerous similarities. Both have been characterised by the growth of the private sector in science and cuts (and shifts) in government funding. Science-industry links and URI patenting have grown in both countries. Furthermore, against neoliberal principle, but consonant with neoliberal practice, the Chinese process has involved a concerted role for the state and an increased blurring of the state/market distinction. This highlights the necessity for active agency of the state in primitive accumulation and thereby indicates the importance of the state apparatus to the US case as well. In contrast to the US, however, the Chinese process has been characterised by the state entering into the market rather than vice versa, reflecting the challenge of Chinese science reform to get commerce into science, not science into commerce.

This takes us to the first of three factors representing singularly important differences between the two processes. All three reflect the absence, in some respect, in China of the structural presuppositions of neoliberal science policy, thereby showing to what extent this political discourse has presumed the universality of the unique American context.

The first issue is the completely different economic structures of the two countries. On the one hand, regarding business, China's science and innovation system has been characterised throughout the reform period by strikingly weak, though improving, business capacity for scientific research. American universities have been able to incubate biotechnology start-ups, for instance, that seek to develop treatments to sell on for development and marketing by 'big pharma' or agri-business TNCs. Conversely, in the absence of these TNCs, Chinese URIs lack a crucial precondition and incentive for profitable spin-offs. This is especially the case regarding the lack of *domestic Chinese* TNCs – in marked contrast to the rise of Japan and South Korea (Nolan 2004) – given the dominance of the state and the importance it places upon national economic development. For while policy reform since 2000 means that the government now considers privately owned Chinese firms potential 'national champions' (Segal 2003), the utter importance of a 'national' firm of *some* description reaping the rewards remains paramount. This also explains the relative lack of privatised genetically modified (GM) research, despite the government's commitment to commercialising GM crops.

On the other hand, regarding science funding, a tacit but crucial presupposition of neoliberal policy in the US was the massive public funding of science in the 30–40 years immediately preceding it, and indeed continuing to the present (see Chapter 4). While the Chinese government also completely dominated science funding until recently, the sums invested were tiny in comparison. Furthermore, the funding of 'basic science' in China has always constituted a much smaller percentage of total R&D funds than in the West, rising to a current peak of less than 6 per cent (OECD 2008). A strong science base there to *be privatised* has thus been largely absent in China, while it has been the basis of the US process.

Finally, and most importantly, lacking a powerful class of finance capital and a strong private enterprise sector dependent upon research-intensive innovation, the *political* impetus of the Gramscian historic bloc behind the American process has been (and looks set to remain) absent. Instead, the Chinese process of privatising science has been *led*, not just implemented, by the central government and in its *own* interests. This substitution of intense and structurally enabled class struggle behind the US process with initially cautious, experimental state leadership of a highly contested reform process is observable in their completely different paces. In the US, a 'counter-revolution' occurred around 1980 (signalled, for example, by the Bayh-Dole Act, the *Diamond v. Chakrabarty* biotechnology decision, the establishment of the CAFC patent court, the explosive emergence of biotech with the record-breaking Genentech initial public offering (IPO) and complete evaporation of moves to legislate recombinant DNA (rDNA) experiments, each of which has its own concrete historical genesis; see Chapter 3) that effected a significant reorientation of the state's regulatory powers towards the interests of the historic bloc, while the privatisation of Chinese science has been a slow and still incomplete process over 30 years of economic reform.

The second difference is thus the pre-eminent importance and leadership of the state. However, the utter dominance of 'national' (or local) interests of development in the process of commercialising Chinese science has by no means produced a process isolated from political controversy, for what *is* the 'national interest' remains complex and contested. A key illustration of this is provided by what, in the US, is a quintessentially neoliberal science, namely GM crops.

Commercialisation is an irreducible aspect of GM-related research. Yet, while in the US research has been utterly dominated by the giant agribusiness TNCs, the organisation of GM research in China clearly shows that the process of commercialising science is filtered through a national project of economic development led by the central government. Furthermore, the Chinese commercialisation of agri-biotechnology also perfectly illustrates how this process has been initiated and dominated by the government, not by a capitalist historic bloc. Political wrangling *within* the government is thus a crucial determinant of what research is pursued and demanded (Zhao and Ho 2005).

In the case of GM research, this political debate hinges on the perception of a dual imperative of commercialising agricultural biotechnology and protecting biosafety, tied together in the government's ambition to modernise China's agriculture. The global context of neoliberalism is reflected in the former. Increasing geopolitical pressures to reap the short-term economic benefit of GM crops have produced growing acceptance across government of GM research and its commercialisation as a matter of strategic national importance (e.g. Ministry of Agriculture 1990). The identification of the goal of national development with the global, neoliberal imaginary of the KBBE is thus apparent, such that agricultural modernisation is often simply equated with GM agriculture.

As a result, agri-biotechnology has been heavily funded, with an exponential increase in governmental funding in the past 20 years. As Huang *et al.* (2002: 675) note, 'government research administrators allocated about 9.2 per cent of the national crop research budget to plant biotechnology in 1999, up from 1.2 per cent in 1986. China's level far exceeds the 2–5 per cent levels of other developing countries'. Given the absolute growth in science budgets, this means that investments in biotechnology research have risen dramatically to $1.2 billion for 2001–5, a 400 per cent increase over the 1996–2000 levels, of which about 10 per cent is devoted to GM rice projects (Jia *et al.* 2004). With an annual outlay of US$115 million, therefore, China leads the world in expenditures on rice biotechnology research (Wang and Johnson 2007).

Research infrastructure has also bloomed, with the establishment of National Key Laboratories (NKL) since the mid-1980s that include 12 working exclusively on, and three with major activities in, agricultural biotechnology (Huang *et al.* 2001, including the spectacular BGI discussed above). Besides NKLs, there are also numerous key 'Biotechnology Laboratories' and programmes under various ministries and local provinces (Huang *et al.* 2004: 3). China is thus arguably developing the largest plant biotechnology capacity outside of North America (Huang *et al.* 2002). Furthermore, all this research has already earned China the reputation as a world leader in the independent decoding of the rice genome and the cultivation of a variety of effective GM strains, including:

> . . . several GM rice varieties that are resistant to the country's major rice pests and diseases, such as the lepidopteran insect stem borer, bacteria blight, rice blast fungus and rice dwarf virus [and . . .] significant progress with drought- and salt-tolerant varieties of GM rice, which have been in field trials since 1998.
>
> (Jia *et al.* 2004: 642)

Yet, even in this classic case of GM agriculture, the process of commercialising Chinese science is hedged and qualified in ways that reveal the central role of the state in this process. First, while commercialisation of GM crop research is an irreducible part of the research programme as a whole, it is 'public', and not 'private', institutions that are orchestrating this process in China. Second, the government has also paid fairly strong attention, in recent years, to the counterbalancing issues of ecological and social risks (Zhao and Ho 2005). The result has been a prolonged prevarication regarding the acceptance of commercialisation of GM crops: even with hundreds of GM varieties waiting in pipeline, the decision to commercialise major GM crops has been pending for 8 years now.[6] Taken together, therefore, these two trends have led to a state-funded public scientific research system in this field in which a presumptively pro-GM (and thus pro-commercialisation) stance is both dominant *and* fundamentally qualified.

Three further issues also exemplify the dominance of the state in the Chinese case. First, the most important, and uniquely Chinese, means to commercialise research has been the state entrepreneurialism of UOTEs. Similarly, URI spin-offs often find it easier to access finance than private start-ups given the good standing of the URI to which they are attached (Chen and Kenney 2007). Even such high profile 'success stories' as Beida Founder or Tsinghua Tongfang continue to bear the name of their university (i.e. state) sponsor. The Chinese state is not merely in business, however, but also deeply involved in the very heart of China's KBBE, i.e. in finance. Venture capital has been dominated by state-owned funds – central or local government or URI – and only recently have private VC funds begun to appear.

This leads to the third issue, namely that even where China has instituted ostensibly neoliberal science policies, they are likely to have different effects, given the alternative political economic context. For instance, the cutting of government funding to URIs did not stimulate patenting and links with private enterprise, as in the US, but the boom in UOTEs. Indeed, even this phenomenon was arguably limited to elite URIs. At most URIs, the lack of resources meant that funding cuts, even when combined with a will to participate in enterprise, were insufficient to stimulate the establishment of UOTEs (Eun *et al.* 2006).

In short, striving for economic catch-up in a global political economic context dominated by neoliberal institutions and imaginaries, China, like other East Asian economies, has been 'torn between continuing mercantilism – now refashioned and recalibrated to put more emphasis on innovation-driving growth – and embracing neoliberal . . . arrangements [to attempt] to benefit from [and/or develop] their own competitive advantages' (Jessop 2001: 34, parentheses added). This conclusion, however, leads to one further, crucial question; namely, how China's qualified embrace of neoliberalism regarding the commercialisation of science will affect China's ability to shift towards strong innovation capacities (and rents) (Wang 2006) and compete effectively in the global KBBE. Certainly, there is no guarantee that China will in fact move 'up the value chain' and shift from a middle- to a high-income country within the global economy (Hung 2009).

This is, of course, a huge subject meriting a concerted discussion of its own to which we cannot do justice in a few short paragraphs. However, a few key considerations may be highlighted here. First, the capabilities in China for hi-tech innovation capable of generating super-rents remain relatively weak (Jakobson 2007, Suttmeier *et al.* 2006). Similarly, regarding the key sector of low-carbon technologies, while some argue that China is 'winning the race', others conclude that there is little evidence of an imminent challenge from China in this sector to incumbent developed economies. For instance, a recent report concluded that intellectual property in low-carbon energy technologies is concentrated overwhelmingly in the large incumbent multinationals from OECD countries, with no significant players from China in the top ten of these sectors (Lee *et al.* 2009). This is partly related to the poor linkages, which underpin technological learning and innovation, between Chinese businesses, both with other domestic firms and with foreign firms present in China through FDI; linkages that have failed to form (in contrast to South Korea and Taiwan, for instance) because of China's fragmented (i.e. provincial and state-owned) industrial structure and financial system (Wang 2006). The lack of VC, discussed above, is also a particularly significant factor here.

The commercialisation (and/or privatisation) of Chinese science, however, may also have a particularly important role in this regard, arising from problems associated with this development; problems that may mark a further important difference with the US case. As we have seen above, a large critical literature, dominated by Mirowski's 'Mertonian Tories', has emerged in recent years regarding the US. This includes concerns about the corruption of the scientific enterprise, the negative alteration of research agendas, the destruction of an open, public sphere of scientific debate and the 'anti-commons' paralysis of research in an unnavigable thicket of private patent rights (Heller and Eisenberg 1998).

Each of these issues raises particular concerns in the Chinese context. It is regarding the last of these issues, however, that the problem of fragmentation and its negative impact on innovation capacities may be exacerbated by commercialisation of Chinese science. The 'anti-commons' involves the proliferation of private interests preventing contribution to the common stock of scientific knowledge that is the basis of a vital scientific culture. In the US, this problem has been discussed (and often romanticised) in terms of the destruction of the pre-existing public sphere of knowledge. In China, however, the reforms *started* with a (different) institutional structure that systematically discouraged collaboration and open discussion of scientific research.

Many commentators and Chinese policymakers seem to hope that the science reform process will unlock a new age of collaboration, based on the repeated arguments of KBBE management literature that innovation hubs need open and serendipitous collaboration. Yet the anti-commons argument is that introducing private enterprise *discourages* free scientific discussion rather than stimulates it. It is thus possible that the imposition of private enterprise incentives onto Chinese science will merely exacerbate its lack of connectivity, thereby harming both Chinese science and its attempts to create a KBBE. Indeed, there is evidence for this problem from the heart of China's research-intensive innovation system, the Zhongguancun science park in Beijing. As Zhou (2005) reports, 'Zhongguancun firms do not cooperate with one another' (see also Guan *et al.* 2005). Yet, it is hard to see how China can develop the innovation capacities needed to become a (let alone *the*) major player in the globalised KBBE without such networks.

Success in the neoliberal KBBE, however, also depends upon the existence of conditions that may be systematically denigrated by neoliberal policy rhetoric but are nevertheless crucial. The commercialisation of China's science also has significant implications as regards these factors. We mention only two here.

The first of these is the demeaning of 'basic science'. The concept of 'basic', hence disinterested or 'blue skies', research is, of course, fuzzy and problematic (e.g. Calvert 2004). Yet, one can acknowledge the irreducible socio-economic aspects of all scientific research without accepting demands that scientific research be appraised primarily or only regarding its *direct* contribution to economic growth. The imposition of such economic managerialism is a hallmark of neoliberal policy, not just for science reform. While China's neoliberalism is generally qualified, however, the economic assessment of science resonates strongly with the domestic tendency to value science overwhelmingly for its economic and technological contributions.

This is epitomised by Deng's 1978 declaration of science as a 'productive force' and the common Chinese (policy) discourse of 'scitech' (*keji*) (Xin 2008), fusing S&T in a way that leaves no place for the former in the absence of its direct economic contribution. It is also apparent in the early and continued identification of China's agricultural modernisation with GM biotechnology. Conversely, Western countries, including the US, have strong scientific constituencies that struggle to uphold the importance of science relatively remote from economic application as an integral component of scientific advance and have powerful normative discourses at their disposal. The particular danger is thus that Chinese science policy actually *adopts* the direct economic assessment of science to an extent greater than any other country. Whether a science system could meaningfully thrive in such circumstances is, at best, an open question. And while the contribution of a vibrant science system to a competitive KBBE is, undoubtedly, much more complex than that posited by the neoliberal conflation with innovation and economic growth, there is little evidence to suggest that the latter two are possible in the absence of the former, especially under contemporary political economic conditions.

Second, many of the critical appraisals of the privatisation of science in the global North have argued that introducing private interests into science undermines not only a central element of the scientific enterprise but also moves to democratise scientific and technological change that are demanded, including for the sake of success in scientific and innovation projects on their own terms, by the social challenges of techno-scientific-intensive late modernity (Beck 1992, Nowotny *et al.* 2002, Wynne and Felt 2007). China, however, does not have a strong public sphere of science in the first place, either to defend the contribution of science to civic and political culture or to demand of science and innovation their broader public engagement. In both cases, these only emerge on the back of concerted political mobilisation. The political culture of China, however, remains relatively closed and cautious, with the omnipresent possibility of a clampdown; another significant difference to other East Asian countries such as Japan or South Korea. Privatising Chinese science thus runs the risk of forestalling the development of a vital and open scientific culture altogether, and hence, hobbling the emergence of (what seem to be) crucial conditions for a competitive KBBE.

5.5 Conclusions

In recent years, Chinese science has been transformed through the introduction of commerce, in parallel to similar developments across the global North, especially in the US. The policy discourse and measures of this process have also assumed an increasingly strong resemblance to US neoliberal prescriptions. These superficial similarities, however, mask strikingly different underlying realities regarding the goals and effects of ostensibly neoliberal policy and the actual changes to the science system. Given the growing global importance and influence of China in science and political economy alike, these differences, in turn, have significant implications for the future parallel trajectories of these two concerns.

Hence, on the one hand, the quantitative and qualitative impact of the attempts to create a globally competitive KBBE in China on the development of Chinese capacities for scientific research are likely to have further and far-reaching implications for the rising geopolitical power of China, played out in both domestic and international socio-political change. Conversely, successes, breakthroughs and continuing frustrations regarding China's place in the global political economy are likely to stimulate growing attempts by China to change the global 'rules of the game'. The transformation of the socio-political model of globalisation would then have significant implications for the political economy of science and innovation in China and the rate at which it does indeed develop scientific (and innovation) capabilities that lead the world. Finally, to all of this we must also add the impact of the 'triple crisis' – of the neoliberal global economy, of science and of ecology – and the multiple and contradictory ways in which they condition both a potential resurgence of and challenges to the (possibly authoritarian) state-led industrial policy (e.g. for a 'green new deal' or 'low-carbon transition', national or global) that still characterises the Chinese political economy.

This analysis also has clear lessons for the standard debate of Economic Whigs versus Mertonian Tories regarding the commercialisation of science. Against the former, it stresses the necessity for coercive state involvement in privatising science, and the dependence of that process on unique structural conditions, thereby challenging the presumptive universality of neoliberal prescriptions. Against the latter, it clearly illustrates that state and market funding of science are not necessarily incompatible, thereby also problematising not just this reified dualism per se in favour of a relational analysis of the state (Gramsci 1971, Jessop 2002a, Poulantzas 1978), but also the simplistic equation of this dualism with scientific virtue and vice respectively. Instead, it forces us to confront the political economic context of *all* science, so that even erstwhile generous state funding must be understood in terms of the particular (e.g. military) interests it served, its position in structures of money-making that justified these funding arrangements and the consequent demands placed upon the science.

Finally, our analysis thereby illustrates how we can only understand the commercialisation of science and its various manifestations – the important similarities of, and connections between, the neoliberal privatisation of American science and the fundamentally qualified Chinese reforms and their equally crucial differences – by situating it in the broader context of political economic changes. As such, this analysis also has significant lessons for STS. For while the politics of science has long been a central concern of STS work, engagement with issues of *political economy* has been lacking, perhaps due to presumed incompatibilities between the agential and constructivist focus of the former and the structural and 'realist' perspective of the latter; issues that we consider in depth in Section IV in Volume 2.

This therefore concludes Section II, in which we have explored some substantive illustrations of the critical and explanatory economics of science. We have seen that the present is an exceptional moment of multiple overlapping and interconnected crises and corresponding emergence of new social realities. The cultural political economy of science and our relations with the institutions of science and scientific knowledge more generally are crucial aspects of both understanding these conjunctures and of working with them towards normatively attractive alternative futures. This is, therefore, to move towards a clear agenda for the response and role of the social sciences as regards the 'triple crisis'. But as a (primarily but not exclusively) epistemic practice, a key question for any such social scientific project is to explain and justify how we got to these conclusions. The first step in reconstructing these arguments, therefore, is some concerted philosophical analysis regarding (scientific) knowledge and its social roles; i.e. questions of epistemology, ontology, methodology and politics. It is at this level that we shall see the singular contributions of critical realism (CR) to the broader project set out here, and it is to this that we now turn.

Further reading

Arrighi, G. (2008) *Adam Smith in Beijing*, London: Verso.
Jakobson, L. (ed.) (2007) *Innovation with Chinese Characteristics*, Basingstoke: Palgrave Macmillan.
Segal, A. (2003) *Digital Dragon*, Ithaca, NY: Cornell University Press.
Sigurdson, J. (2005) *Technological Superpower China*, Cheltenham: Edward Elgar.

SECTION III

Critical realism and the importance of ontological attention

6

TOWARDS A CRITICAL REALIST ECONOMICS

Refinements to 'realism'

6.1 Introduction

In Section II, we explored recent substantive developments in the political economy of science, especially regarding its progressive commercialisation. In doing so, we illustrated the insights available concerning our current predicament from a cultural political economy perspective, employing a relational Marxism. It remains, however, for this perspective to be justified in theoretical terms, regardless of the substantive lessons we have already seen it to be capable of yielding. This theoretical argument will occupy us for the remainder of the book, setting out the steps from the mainstream economics of science to a cultural political economy of research and innovation (R&I), while in the process highlighting the theoretical insights from the (inchoate and ongoing) synthesis and concerted engagement of political economy and science and technology studies (STS).

Underpinning this mutual accommodation and challenge, however, are a series of fundamentally philosophical questions regarding issues of ontology, epistemology and methodology that are insistently prominent for an economics of science. This is not just because of the epistemological task of theoretical justification just described, though that is certainly crucial, but also because a credible alternative to the economics of science must also grapple with the nature of 'science' for there to be an economics *of*, the nature of the economy and social reality more generally in which science and scientific knowledge are embedded, and their interrelations. Furthermore, beneath all of these are the most abstract questions of ontology, regarding the nature of reality per se; what it is to be 'real'. Our starting place (in Section III) is thus an examination of these philosophical questions, introducing the perspective assumed, advocated and developed in Sections IV and V in Volume 2.

In order to see why we take this approach, consider again the critical context of a mainstream economics of science. We have seen that the actual contribution of mainstream economics to an economics of science is at best highly circumscribed and that, certainly, it cannot be the basis for a comprehensive research programme on such issues given the incompatibility of some of its fundamental tenets regarding the workings of markets and the nature of scientific knowledge (Mirowski 2009). We may be tempted, therefore, to forsake mainstream economics altogether when searching to construct an economics of science and move on to other, more promising perspectives. That is certainly the approach that is taken in this book. The reasons for doing so, however, extend beyond the intractable analytical problems identified by Mirowski. In particular, two further problems with mainstream economics add to the motivation to leave mainstream economics behind.

In both cases, the politics of mainstream economics is a key concern. First, on the one hand, mainstream economics, assuming the garb of a social physics, modelled on a positivist (mis)understanding of the nature of science (see pp. 150–152), not only asserts the political neutrality of its analysis but does so proudly as a badge of its scientific status. The close interaction of economic and political matters, however, as most obviously in the power associated with command or ownership of economic resources, immediately shows the untenability of such an artificial separation of issues and its associated division of cognitive labour. And, furthermore, it illustrates how the apolitical posture of mainstream economics simply forces its political commitments and implications into the shadows.

Second, the full force of the error of mainstream economics denying the relevance of power to its subject matter is most graphically demonstrated in the performative contradiction of its own moves as a discipline to take over all alternative forms of explanation and subject matters across not only the social sciences, but also increasingly the psychological and biological; what Ben Fine (1999, Fine and Milonakis 2009), inter alia, describes as 'economics imperialism'. The utter failure of economics as a discipline to predict, let alone mitigate, the great financial crash of 2008 and the continuing fallout of the deepest economic recession since the 1930s has lead to some soul-searching (e.g. *The Economist* 2009b) and, hence, one may suppose, a certain stumbling in the progress of economics imperialism. But there can be no expectation that this will lead to a fundamental break with the canons of mainstream economics, no matter how discredited it is intellectually (even in the eyes of economists themselves (e.g. Harcourt 2010)); if anything, the evidence is quite the contrary (e.g. Davies 2010). Indeed, as Fine shows (2010), mainstream economics has been a cannibalistic living-dead 'zombie' science for decades: dead as regards its intellectual vacuity; living as regards its continuing utter dominance of the social sciences; and cannibalistic in its insatiable appetite to consume and destroy alternative research programmes.

While economics imperialism is undeniably a serious problem, for an economics of science especially it does at least demand that the assault of mainstream economics to heterodox or alternative modes of explanation be confronted head one. For mainstream economics not only cannot provide an insightful economics of science but it also wants to prevent other approaches from doing so, ruling them out as 'scientifically' illegitimate. In short, given the intellectual and political dominance of mainstream economics, our first task has to be to justify the wholesale move away from this perspective and our focus on other perspectives. In responding to this challenge, however, we will not simply counterpose the tacit politics of mainstream economics to those of other perspectives that we (may or may not) find more agreeable. Such an argumentative strategy is bound to lead only to a self-undermining judgemental relativism that concludes, at best, with the agreement to disagree, while leaving mainstream economics and its dominance intact.

Instead, an alternative strategy not prone to this problem (though not immune to others – see 7.4, pp. 172–178) is to turn to questions of philosophy in order to demonstrate that, far from standing on the epistemic higher ground of a unique claim to scientific justification, mainstream economics is itself without epistemic warrant. In short, we can seek instead the determinate negation of this dominant framework by altogether uprooting it. As Fine's discussion of 'zombieconomics' makes clear, this cannot license any expectation of such a critique alone producing determinate negation *in practice*, but we may at least consider it a necessary if not a sufficient condition of this change in the discipline.

In taking this approach, the particular argument that we will deploy and build upon is that of Tony Lawson and the broader philosophy of science called 'critical realism' (Archer *et al.* 1998). As Fullbrook (2008) notes, Lawson's criticism of mainstream economics in the past 15 years has fundamentally changed the debate within the philosophy and methodology of economics (if not its practice) by insisting upon the need for explicit consideration of ontological presuppositions, our understanding of the nature of reality that is always already assumed; what I am going to call 'ontological attention' in contrast to the problem that Lawson (1997: 37) identifies as 'ontological neglect', the central (and unique) concern of critical realism (henceforth, 'CR'). Furthermore, following CR, it is argued here that the way to examine such ontological presuppositions is by way of the particularly philosophical form of reasoning, the transcendental argument (TA) (see Chapter 8). As such, CR proposes a complementary but non-foundationalist division of intellectual labour between philosophy (especially regarding ontological issues and their epistemic repercussions) and social science that also, contingently, affords the determinate negation of mainstream economics and its positivist philosophical pretensions on the grounds that it is an insufficiently realist, and so judgementally ungrounded, research programme.

In articulating the critical realist critique of mainstream economics and its suggested alternative model for a scientific economics (or an 'economics as social theory' per Lawson (1997, 2003)), however, we must also deal with the key questions of 'why CR?' and 'what does it have to offer?' This demands two further steps, roughly reflecting the two questions of 'what is realism?' and 'what are the methodological implications of this argument for the social sciences?' First, not least to tackle persistent and widespread misunderstanding of critical realist argument (including Lawson's), we must clarify the important difference in the sense of the (notoriously polysemic) term 'realism' as used by CR as opposed to mainstream debate in the philosophy of science; in brief, an ontological realism rather than an epistemological realism (cf. Sismondo 2007). This difference hinges on the unique concern of CR to pay *explicit attention to ontological presuppositions*.

But this difference in meaning of 'realism' is not simply a matter of different conversations crossing likes ships in the night or alternatives that may be mutually compatible as totally orthogonal. Rather, the critical realist argument's focus on ontological issues has significant implications for the epistemological debate that challenge the relevance of the dominant understanding of the term 'realism'. In short, by examining the ontological presuppositions of particularly significant social practices, our ontological understanding is gradually spelled out and transformed, thereby clarifying what we, always already, do in fact – correctly or incorrectly, but socio-historically specifically in each and any case – take to be the nature of reality. But since the starting point of a realist perspective is that knowledge must 'fit' to being and not vice versa, it follows that changing our understanding of the nature of reality is also thereby to transform our understanding of what is substantially required of the realist science studying that reality. Accordingly, the following argument offers a series of refinements to our understanding of what a 'realist social science' would be.

Following on from this opening to revised methodological prescriptions for a realist social science, however, the second step (in Chapter 7) is to confront a number of important criticisms of Lawson's argument, in particular regarding his positive reconstruction of a critical realist economics. We will consider two lines of argument: first, by way of immanent critique; and second, from other broadly constructivist traditions that also employ post-positivist philosophies of science. The former leads to the proposal of a critical methodology in contrast to Lawson's contrastive methodology; a project that is then taken up and fleshed out in subsequent chapters as a critical, explanatory political economy of science. The latter opens up an ongoing dialogue between CR and other post-positivist perspectives that, crucially, instantiates the promise of a *mutually* informative engagement between philosophical and social scientific research programmes that is both explicitly essential to the non-foundationalist programme of CR and yet inadequately taken up in the critical realist literature. This dialogue also continues throughout the rest of the book, illustrating a productive tension that serves to strengthen and justify a CR approach rather than undermine it, but by way of qualifying and relativising CR arguments.

6.2 Refinement 1: From ontological attention to transcendental realism

Our first task is to articulate the argument made by CR for the importance of ontological attention and hence (a particular conception of) realism in science. This argument starts from the observation that it is a necessary condition of the intelligibility of each and every theory of knowledge (and thus each and every actual scientific theory) that reality is such that it can be thus described (e.g. Lawson 1997: 19).[1] Furthermore, all theories are thus 'realist' in the sense of presupposing *some* conception of the nature of reality (ibid.: 19, 48), i.e. *all* theories, and indeed all intentional actions that employ belief, have ontological presuppositions. We are therefore bound, simply as a matter of consistency between our beliefs and our practice (including between our explicit philosophical opinions and the ontology actually presupposed by our scientific ones – our (scientific) beliefs themselves being a form of 'doing' in the world), to admit that we are committed to particular understandings of the nature of reality and that these must also be uncovered and tested for consistency.

Examination of these ontological presuppositions is a strictly philosophical task, employing transcendental reasoning, which asks the question 'what must be the case given that the premise (of the transcendental argument) is intelligible?' (see Chapter 8; Tyfield 2007). This is the primary focus and distinguishing feature of CR as a philosophical project: to counsel the need for such explicit consideration of our ontological presuppositions, or 'ontological attention'. Such a 'critical' philosophy can work by immanent critique, provided by the critical context of generally accepted premises (which gain both their particular meaning and ontological purchase, through being genuinely believed to be true, from this socio-historically specific and pragmatically given critical context), and transcendental argument to expose underlying contradictions in our given understanding so that there is an irreducible role for philosophy (qua ontology or, more pejoratively, 'metaphysics') as 'underlabourer' for the sciences (Lawson 1997: 45). As such, CR marks a whole-sale break with much (if not most) modern mainstream philosophy of science in its turning *back* to ontology as the fundamental and distinctive task of philosophical argument, against the anti-metaphysical programme of the former.

Taking these points together, then, we see immediately how the concerns and purposes accorded to realism and philosophy by CR are substantially different to those of the mainstream philosophy of science debate concerning 'realism'. The latter is primarily interested in the epistemic question of whether it is right or wrong to treat mature scientific theories as more or less true (Psillos 1999: xvii) while the former is primarily interested in the practical differences, including to social scientific work, made through considering explicitly what we do already necessarily believe, albeit often tacitly, to be the case about reality. For CR, therefore, the philosophical question arises in the context of an attempt to *do* something with the answer and of the need to address this question in order to make further progress with our given practical problem. Conversely, mainstream philosophical debate generally takes place in the *absence* of specific (social) scientific interests and is conducted for its own ends. Indeed, the practical context of philosophy on the CR conception is a major difference that resonates through its argument, in particular sponsoring the critical conception of philosophy (and social science, as argued below) for which it argues. It may also be argued that its privileging of ontological over epistemological questions makes CR distinctively consistent in its realism.

This is important because these are not merely orthogonal debates that, unfortunately, confusingly use the same term. Rather, CR also leads to some substantive conclusions regarding the nature of reality that also entail some significant resituating of the purely epistemological argument of the mainstream debate. The first conclusion is the inescapable presupposition of a mind-independent reality per se, i.e. not conjured up in and by the act of thinking it (an ontological realism as opposed to an ontological idealism/irrealism, or what may be classified as a 'realism^{ID}'). But simply admitting realism^{ID} explicitly (proclaiming oneself a realist^{ID}) is itself insufficient if our goal is to attain consistency between our beliefs and our actions, because the very fact that all beliefs have ontological presuppositions entails that these must also be compatible with this conclusion. By making the ontological presuppositions of our express beliefs themselves explicit, therefore, we can conclusively reject those beliefs whose presuppositions are incompatible with realism^{ID}.[2]

Nor does this process of comparison end there because realism^{ID} is simply the most fundamental, and so the least discriminating, form of realism. Theory/practice consistency also depends, therefore, on the explicit consideration of the ontological presuppositions of particular social practices in order to articulate in greater detail what reality must be like given that these are possible. As this elucidates the nature of reality already immanent in our understanding, this makes a crucial contribution to the (always – see Chapter 7) ongoing process of seeking consistency, by way of such comparisons, between our beliefs (explicit and implicit) and our actions. Finally, given that knowledge must 'fit' to being, it follows that clarifying our understanding of the nature of *reality* through each stage of this ontological investigation is also simultaneously to clarify our understanding of the meaning of '*realism*' as regards (social) science.

One particularly important contribution to this learning process is the ontological analysis of a social practice that, as a matter of critical context, we accept gives us knowledge of reality. For it is a necessary condition of the intelligibility of the fact that this practice produces knowledge that reality is such that it can come to be known in that way. In pursuing this line of argument, therefore, we can start from the general acceptance (including among philosophers of science – hence a critical and socio-historically specific but appropriate context) that experiments are the archetypal scientific practice accorded epistemic warrant (e.g. for the 'testing' of theories), i.e. that these accepted paradigmatic examples of science *intelligibly* produce knowledge (Bhaskar 1998, 2008). Bhaskar's seminal contribution to the philosophy of science rests on this exceptional philosophical innovation of subjecting actual scientific practice to ontological attention. In examining the ontological presuppositions of experimental practice, CR argues that we are furnished with an ontology that substantially alters our understanding of the nature of reality and thus of realism.

Let us consider this argument. An experiment is a set-up controlled by the scientist in order to examine the particular correlation between distinct events in these closed conditions. It follows that the intelligibility of experimental practice depends upon the fact that such event regularities would not otherwise be in evidence. Yet, for this to be possible it cannot be the case that the causal law identified in this way by the scientist exists at the level of events, for such 'laws' qua event regularities are created by her intervention. In short, 'a *real* distinction between the objects of experimental investigation, such as causal laws, and patterns of events is thus a condition of the intelligibility of experimental activity' (Bhaskar 1998: 9). Furthermore, if the commonplace application of causal laws, identified in this way, outside the laboratory is to be intelligible:

> causal laws must be analysed as the tendencies of things, which may be possessed unexercised and exercised unrealized, just as they may of course be realized unperceived (or undetected) by people. Thus in citing a law one is referring to the transfactual activity of mechanism, that is, to their activity as such, not making a claim about the actual outcome (which will in general be co-determined by the activity of other mechanisms).
>
> (ibid.)

The ontological presuppositions of scientific experimental practice, therefore, are for a 'transcendental realism' (as opposed to an 'empirical realism'), in which there is a real 'ontological gap' between domains of the empirical, actual and 'real' (or, metaphorically, 'deep'; Fleetwood 2001: 67), each term being merely a subset of those that follow.[3] While the first of these three terms is unchanged by this analysis, as that which is the object of sense experience, the latter two terms have undergone a subtle redefinition in this process. As regards the third term, the real stratification of reality into these domains entails that reality is understood greatly to exceed its particular manifestation at any one time. But since actuality is now to be understood in contrast to the 'deep' or potentiality of reality, actuality refers merely to the surface of actual things, states of affairs and events.[4]

Transcendental realism can therefore be contrasted with an 'empirical realism', which is not only the default position of modern anti-metaphysical philosophies of science (especially of various 'positivisms' and their progeny) but also of the 'common sense' that these often set themselves the task of preserving (viz. 'if I can bite it, it's real'). This is a presupposed ontology that eschews such a distinction of realms or domains and collapses all three to the same level of actual things or states of affairs (Lawson 1997: 19–20). For an empirical realist ontology there is no distinction between what 'is' and what is actual. An empirical realist ontology also sponsors particular understandings of key concepts associated with issues of ontology, such as causality, and derivatively, of knowledge and science. In particular, because reality is confined to the level of events, the only way to achieve scientific knowledge about it is through the identification of patterns of regularities or correlations between such events.

Conversely, transcendental realism elicits a substantial change to our under-standing of these terms. We have already seen one example in the neologism of 'transfactual' activity of that which is real. But further implications of this ontological redefinition include, for instance: acknowledging a necessary 'intransitive' dimension to science, namely the presupposition of a real (relatively subject-independent) object of study or 'ontological realism'; a redefinition of causes as real causal powers and their analysis in terms of tendencies not *ceteris paribus* correlations of events; the stratified emergence of these real causal powers, licensing a 'synchronic emergent powers materialism' (SEPM) that rules out an eliminative reductionistic mode of explanation; a relational ontology of emergent phenomena that are constitutively relational, characterised by dynamic internal and external relations; accepting that science uses both a perceptual *and a causal* criterion for the ascription of reality to a theoretical phenomenon; and the identification of the central move in science as being that from events to identification of the causal mechanisms underlying them, thus making the central form of scientific inference neither deduction (the movement from general laws to particular instances) nor induction (vice versa) but 'retroduction'.[5]

Furthermore, the ontological gap between actual and real and the a posteriori observation of the 'openness' of reality to change entails that there can be no expectation that the real causal powers will manifest in event regularities at the level of the actual. But the priority of being over knowledge demands that a science must identify the real causal processes involved in a particular episode or phenomenon. Accordingly, the criterion of explanatory power, defined in terms of (the broad range of) the insights a theory offers into these real causal powers, and *not* predictive success, is the major criterion of theory assessment. Science is thus accorded a primarily *explanatory* role. The unfortunate conflation in much philosophy of science of explanatory and predictive success as regards the physical sciences (the *locus classicus* being the DN model of Hempel (1962)) is then explained as dependent on the exceptional nature of the objects in this case, including the relatively stable and enduring nature of the causes identified. Similarly, the singular importance of experiments to the success of the (laboratory) natural sciences depends precisely upon the *nature of the subject matter* being such that experiments upon it afford access to (what may be socially confirmed as) real underlying causal mechanisms through prediction in artificially closed conditions.

A further important amendment to our understanding of science arises from the need, given the current state of debate in the philosophy of science, to reckon with the phenomenon of scientific change. Given the intransitive dimension, one can see how such scientific change is intelligible, for our knowledge may develop, improving in its 'capturing' of reality as conceptual elaboration takes place. But in order to avoid the 'absurdity of the assumption of the production of such knowledge *ex nihilo*, it must depend upon the employment of antecedently existing cognitive materials' (Bhaskar 1998: 11). Since such cognitive materials (as well as technological and institutional conditions) of scientific practice are always socio-historically specific – the product of particular, ongoing and open historical trajectories – there is thus an irreducible 'transitive' dimension to science too. This gives rise, therefore, to an epistemic relativism, in which knowledge is always situated and hence relative to its particular context. This is not, however, a *judgemental* relativism because, given ontological realism, reasoned discrimination between competing theories on the basis of their explanatory power remains possible, hence affording a provisional and fallible judgemental rationality (e.g. Lewis 2004b: 11). The philosophical chimera of absolute (or 'God's eye' or 'view from nowhere' (Nagel 1989)) knowledge is thus willingly abandoned as both empty and unnecessary.

Finally, what are the implications for our understanding of realism? One in particular stands out, turning on the original question of whether our understanding is compatible with the existence of a mind-independent reality. For this is not the case so long as our presupposed understanding of the nature of being reduces reality to that which is empirical, because such a reality is not, by definition, independent of human experience. It follows that only by explicit adoption of a *transcendental realist* ontology – one that separates the domains of the empirical < actual < real – can such a realism[ID] be consistently held, i.e. not only as regards explicit belief but also regarding whether or not the presuppositions of our other beliefs and theories are compatible with this explicit belief. Furthermore, we have already seen that all theories (and all intentional agency) have the primary ontological commitment that there is a mind-independent reality. As a simple matter of theory/practice consistency, therefore, we must be not only self-consciously realist[ID] but also transcendental realist (e.g. Bhaskar 2008: 26, Lawson 1997: 51). In short, and following CR, the first refinement to our understanding of what it is to be a realist scientific enterprise is that we must adopt an explicit philosophical ontology of transcendental realism and the altered understanding of the goals of science to which this gives rise (but see 7.4, pp. 172–178).

6.3 Refinement 2: Realism and social ontology in economics

So far we have examined the critical realist argument of the need for explicit consideration of our ontological presuppositions as regards our understanding of the nature of reality per se and, derivatively, of science. We now turn to consideration of the *social* sciences and economics in particular; our primary concern. In doing so, we will consider the argument of Tony Lawson regarding the implications for economics of the critical realist argument for ontological attention and thereby provide the second refinement to our understanding of the nature of realism.[6]

Lawson's argument regarding economics starts from the premise that our knowledge or theories concerning social reality must be such that they are compatible with the nature of their subject matter, as argued above. In order to provide this comparison for economics, Lawson examines the ontological presuppositions of, on the one hand, mainstream economics and, on the other, the social phenomenon of human choice, selected for reasons of critical context (as explained above) given that it is something that economics generally emphasises. By comparing these ontological conclusions, he concludes that mainstream economics is inadequate because its methodology is incompatible with the nature of its subject matter.

Taking these in reverse order, Lawson offers a transcendental argument from the premise of human choice in order to examine the ontological presuppositions of this central feature of its subject of study (Lawson 1997: 30 ff.). Choice presupposes, first, the ability of a person to have acted otherwise than they did. Economic reality, therefore, cannot be such that event regularities will in general hold because any such regularity can always break down through alternative choices being made. Furthermore, the post hoc failure of econometrics to identify significant economic event regularities provides further evidence of their absence. Second, choice also presupposes intentionality as regards actually *choosing* different courses of action. For such intentional choice to be possible, however, there must be relatively enduring objects of the knowledge employed in reaching such a decision, the material conditions (in the Aristotelian sense) upon which efficient agency may be employed. Yet, we have just seen that these objects of knowledge cannot be event regularities.

Indeed, these material conditions can be shown to be specifically social structures that exist at the ontological level of the non-actual real. First, given that the difference between the natural and social sciences is that the objects of the latter are themselves intentional, 'social' is here defined as just such a dependency on human intentional agency. Second, given the *causal* criterion of ascription of reality discussed above, it can be readily shown that only specifically *social* causes, while not themselves directly perceptible, render intelligible certain physical states of affairs that *are* directly perceptible. Thus, for instance, the physical interaction of a team game is unintelligible in the absence of the specifically social structure of the *rules* of that game. Similarly, as regards the economy, the explosion over the past centuries in the physical movement of goods and people around the globe is unintelligible in the absence of the social rules of exchange. Such social structures, however, are not existentially independent 'things' because they only exist insofar as there is intentional agency for them to condition. Since they are not actual and directly observable things, it follows that the relevant (material) causes must lie at the level of the domain of the non-actual real or 'deep', just as with natural science.

These social structures (rules, relations, roles and practices) are therefore the subject matter of a realist social science. Furthermore, as real but non-actual, the ontological hiatus between real and actual, coupled with the genuine ability of an intentional agent to choose alternative courses of action, together entail that these social structures are most unlikely to manifest spontaneously in event regularities that are significant. It follows that the second refinement to our understanding of realism as regards the social sciences is that a realist theory must use a methodology the presuppositions of which are compatible with a social ontology of real, non-actual and developing social relations. Conversely, a theoretical or methodological approach that can only offer knowledge of event regularities is of little use in attaining knowledge of the economy.

Furthermore, CR argues that the methodological option employed by the natural sciences in order to navigate between the 'real' subject matter and 'actual' events that can afford epistemic access to them – namely experiments – is foreclosed for the social sciences (Bhaskar 1998: 45, Fleetwood 2001: 68, Lawson 1997: 195). Recall that the goal of a social science is to attain knowledge of the non-actual real social structures in a particular social context of interest. Yet, as non-actual, these only ever constrain/enable (that is, *condition* or *mediate*) and *never determine* human action. Rather, a social structure is existentially dependent on intentional agency, which is the ability always to choose to do otherwise. It follows that the various closure conditions necessary for experimental access to real structures from event regularities cannot be met for social phenomena and that a theoretical framework that depends on event regularities is also denied this possibility for establishing its epistemic efficacy.

Turning then to the obverse side of the comparison, Lawson argues that the ontological presuppositions of mainstream economics are such that it is only capable of knowledge of event regularities. First, he offers a characterisation of mainstream economics as essentially concerned with mathematical modelling. In this, he is by no means alone, with even Nobel-winning economists noting and bemoaning (while others celebrate) the dominance of maths over the mainstream of the discipline.[7] But Lawson takes a controversial step at this point, arguing that this modelling methodology is 'deductivist'. By this he means it employs 'a type of explanation in which regularities of the form "whenever event x then event y" (or stochastic near equivalents) are a necessary condition' for the intelligibility of statements about that subject matter in that particular form (Lawson 2003: 5, also Lawson 1997: 17, 1999b: 4). It may be admitted that Lawson does not use the term in the most obvious sense, i.e. as concerned with deduction from axioms (Hands 2001: 323, note 35, Hodgson 2004: 4), but his definition of the term clearly is related to this meaning because the purpose of employing and postulating such event regularities is usually to provide a covering law for deductive explanatory purposes.

Lawson's particular formulation of 'deductivism', however, arises from his primary interest in examining the ontological presuppositions of mainstream economics as a body of thought; this, in turn, directing his attention for a definition of mainstream economics away from its substantive concerns to the particular *forms* of explanation it employs. Hence, he stresses that he is interested in the 'form' or 'structure' of deductivist explanations, which presuppose 'closed' systems, defined as those in which such event regularities occur (Lawson 1999b: 4, 2003: 13). But given his focus on the form of explanation and its ontological presuppositions (i.e. what reality must be like if it can intelligibly be understood in that form), other factors such as whether the variables in the model are fictitious or claimed to be realistic, or whether 'an inductive or . . . deductive emphasis is taken' (Lawson 2003: 5) are entirely irrelevant to his analysis. None of these factors alter the form of explanation adopted and the relevant ontological presuppositions follow from this alone.

In what sense are mathematical models necessarily deductivist? Such mathematical models necessarily take a form of formulae in which the setting of certain figures (the exogenous variables) invariably sets the value of the others. But this is precisely for the explanation to take the form that 'whenever X then Y *ceteris paribus*', because setting the exogenous variables (X) will *always and without fail* produce the same actual result (Y). Furthermore, it is a condition of the intelligibility of this form of explanation being able to provide knowledge of its subject matter that the nature of this object is such that it does afford such event regularities. Conversely, if the subject matter does not avail of event regularities than it cannot be known by a body of knowledge that takes this form.

Finally, then, drawing the two sides of the argument together, since the nature of the economy is such that it does not in general license such event regularities, an economics devoted only to mathematical modelling is radically incapable of offering knowledge about the economy. This therefore explains (at least in part) why econometrics has had so little post hoc success in isolating significant event regularities in the economy. To be sure, as Lawson's argument has unfolded over time, it has been made clear that such mathematical skills are not completely redundant but in fact extremely important, with substantial argument *within* CR over how and when it is epistemically permissible to use them (Downward 2004). But it remains the case that a mathematical economics could not itself answer this question, which instead depends on a substantially different form of economics, an 'economics as social theory' (Lawson 1997), that would provide the qualitative analysis necessary to justify the skilful and reasoned use of quantitative analysis. There is thus the need for guidance regarding the methodological implications of realism for this prior economics.

6.4 Clarifying Lawson's argument: Criticisms from economic methodology

Lawson therefore offers a far-reaching critique of the current practice of mainstream economics. But given his conclusion of the need for a major 'reorientation' of economics (Lawson 2003), it is no surprise to find that his argument has received considerable criticism (Fullbrook 2008). Many such criticisms, however, are simply talking past Lawson, addressing arguments that he does not make. Given the widespread misunderstanding of Lawson's argument that this evidences, let us focus on several criticisms that resonate through the literature in order to clarify what he is and is not arguing. This serves two purposes in particular: to show that the focus of Lawson's argument is simply that a theory's ontological presuppositions must be compatible with the nature of its subject matter, thus simply applying the CR focus of ontological attention to economics; and that his criticism of mainstream economics is therefore justified, thereby setting up the requirement for an alternative methodology:

(1) First, it is argued that Lawson's characterisation of mainstream economics as 'deductivist' is wrong so that his criticisms are directed at a straw man. For instance, Hands (1999) argues that mainstream economics is only plausibly cast as deductivist in the sense of seeking event regularities if one swallows its pious methodological pronouncements.[8] Conversely, if (as one would expect of a realist) we examine actual economic *practice* as exemplified in economics textbooks and papers, we find no such event regularities at work, or at least hardly any.

(2) Second, it is argued that Lawson is wrong to identify the main problem with economics as a lack of realism. For instance, Hausman (1998, 1999) draws the distinction between various kinds of realism and argues that two types are possibly relevant to economic methodology: realism versus instrumentalism (realism[I]) as the goal of science; and realism versus (epistemic) anti-realism regarding the truth of unobservable theoretical entities, 'scientific realism' (realism[S]). He then argues that most, if not all, serious commentators in current economic methodology are realists[I] so that Lawson is distinctively 'realist' only in the realist[S] sense but that realism[S] is irrelevant to economics. For economics employs theoretical terminology that arises directly out of social life so that one cannot question the (observability and hence) *existence* (as opposed to the *nature*) of these entities without also being a radical sceptic in everyday life. It follows that labelling Lawson's methodological program as 'realist' is both otiose, for this is not a distinguishing feature of his argument, and misleading, because it suggests that it is distinctive and so distracts attention from what is *really* unique to his position, namely his 'ambitious theorising' and 'controversial metaphysics' (Hausman 1998: 204–205).

(3) This leads directly to the third argument, also from Hausman, that Lawson presents a false dichotomy argument to economists and economic methodologists, namely, 'Either they can accept a view of science as exclusively the search for exceptionless regularities among observable events (glorified correlation-spotting), or they can accept critical realism' (ibid.: 204). Indeed, it is no exaggeration to say that this dichotomy is the major point of contention within economic methodology regarding CR.

In responding to these various criticisms, a common thread that runs through each of them is a misunderstanding of Lawson's argument on the part of the critic revolving, in particular, around the conflation of express or avowed philosophical position regarding realism and (probably tacit) presupposed ontology. I stress that this is not to dispute or impugn the realist sentiments of Hausman, Hands, Vromen, etc. but simply to point out that these realist beliefs *cannot be consistently held* without revising their ontological commitments, i.e. without changing how they understand such terms as 'cause', 'reality' – as well as, for an economics, 'social reality' – and derivatively 'science'. But this is precisely the central claim of Lawson's argument so that, in overlooking this, such critics are simply not engaging with it.

Turning to the specific criticisms then:

(1) First, Hands is simply wrong to argue that there are no event regularities in mainstream economics, for even the examples he outlines in order to disprove Lawson are, quite the contrary, perfect examples of just the event regularities that Lawson identifies. For instance, Hands examines a 'canonical' textbook (Arrow and Hahn 1971) in order to see whether or not event regularities are as important to mainstream economics as Lawson claims. Having searched in vain for any in the first two chapters, he finds the following scenario explained in Chapter 3:

single-output firms, with knowledge of all technically possible relationships between their output and all the various combinations of inputs that
they could possibly have access to, engage in timeless . . . production . . . These
firms maximize a continuous profit function defined over a bounded and
strictly convex production set that admits free disposal. These firms are shown
to generate an aggregate supply correspondence that is continuous.

(Hands 1999: 177)

But once again, Hands argues, there is no event regularity in evidence here and
so on throughout the book. He concludes, therefore, that Lawson's 'deductivist'
characterisation of mainstream economics is simply wrong and thus his critique
fails.

Conversely, far from demonstrating the absence of event regularities, this
description of Chapter 3 of the Arrow and Hahn textbook provides a perfect
illustration of the kind of event regularity argument Lawson alleges are ubiquitous
in current mainstream economics, namely that whenever such firms 'maximise a
continuous profit function defined over a bounded and strictly convex production
set that admits free disposal' they 'generate an aggregate supply correspondence that is continuous' or 'if X then Y *ceteris paribus*'. Why then does Hands say
otherwise? It is clear that he is taking Lawson's argument regarding the importance
of event regularities to such economics to mean that the economic framework
takes itself to be warranted only if it provides exceptionless correlations between
phenomenal sense data, i.e. 'event' is understood in a strictly empiricist sense. Hence
he argues, regarding the above example, 'It is not exactly clear what status such a
firm might have in a positivist world where the only meaningful propositions are
those involving sense experiences or the purely analytic propositions of logic and
mathematics. . .'. (ibid.: 177).

But not only is this to recast Lawson's argument in such a way that it is patently
absurd as a characterisation of economics, it is also to confuse his argument that
such economics *presupposes* an empirical realist ontology with the altogether
different argument that it expressly conducts itself (let alone makes pious methodological statements) in strictly empiricist terms. But these are entirely separate
points: empirical realism is a matter of (tacitly) presupposed understanding of the
nature of being per se, while empiricism is an explicit epistemological position,
and Lawson is concerned only with the former. As discussed above, the 'empirical
realism' of 'event regularities' that Lawson describes refers to an understanding of
the nature of reality presupposed by the form of explanation adopted. As such, the
'events' correlated need not just be 'events' qua empirically sensed instants, but
equally can be things, classes, sets, properties, etc. and can be actual, counterfactual
or entirely fictitious. They can even be called 'capacities' or 'structures', seemingly
adopting a critical realist ontological perspective (Lawson 1999c: 224). In short, it
does not matter what the 'status' of the Xs and Ys are but only that they are related
in the form of a regularity. It follows that Lawson's argument readily incorporates the type of explanation outlined by Hands as mistaken counterexample

(see Fleetwood 2001, Lawson 1999c: 223 ff., 2003: 13) and that his critique of mainstream economics stands.

(2) We turn next to the relevance of 'realism' to the problems of economics. First, given the importance of such unobservables to the economic theory articulated and used in later chapters, we may note with Lawson (1999a) that Hausman, in his rejection of the importance of unobservables to economics, has simply assumed that 'economics' means only the mainstream. For even the most casual acquaintance with heterodox traditions will show that these do employ unobservable theoretical entities and as *central* to their theoretical frameworks: e.g. (possibly long-wave) business cycles, technoeconomic paradigms, surplus value, social institutions, etc. Indeed, regarding Marxian economics, Hausman (1998: 200) expressly admits as much, but is simply dismissive of this having any significance for his argument: it is effectively *not* economics.

What is most important for present purposes, however, is that, like Hands, Hausman has completely misread Lawson's argument. For Lawson's position is *not* primarily 'realist' in either of the senses that he considers, so that his whole argument is directed at a straw man. As we have seen, the CR argument is primarily realist in an altogether more fundamental way, namely that there is a mind-independent reality for science to be 'of', i.e. realism[ID]. 'But surely', Hausman and critics may respond, '*none* of us denies this'. This, however, is once again to conflate explicit philosophical allegiance with implicit ontological presupposition – and thus precisely to overlook Lawson's basic point: that because we are inevitably realist[ID] in practice we must attend to our ontological presuppositions in order to ensure they are compatible with this. An argument for realism[ID], therefore, is only uninteresting or unnecessary for current methodological debate so long as ontological commitments are expressly examined and revised to accommodate such a mind-independent reality and, to repeat, *this means transcendental realism*: the acceptance that reality is simply (much!) greater than our knowledge and experience of it. Moreover, only once this first argument is admitted can Lawson's subsequent discussion of social ontology be understood, because the *nature* of the argument is the same in both cases.

(3) Finally, the same misunderstanding is in evidence as regards the third objection. As presented by Hausman, one can see that it is indeed a false dichotomy because he interprets one side as *explicitly* engaging in correlation spotting and the other (possibly) as the full methodological argument of critical realism. Conversely, the dichotomy Lawson presents is that of a *presupposed ontology* that is empirical realist and one that is transcendental realist. But, as Lawson (1999a) himself has argued, there is nothing false in *this* dichotomy, for either one explicitly examines ontological commitments, and hence admits an ontology that admits of a trans-human reality, or one does *not* and so remains bound to a presupposed understanding of the nature of being that does *not* admit this. Coupled with the central importance of the type of realism captured by this distinction, as argued above,

it is perfectly justifiable to present such a dichotomy as an exceptionally significant one as regards the 'realism' of a particular theoretical or methodological standpoint.

In short, then, in each case these criticisms are based on a failure to acknowledge Lawson's central argument: that there is a distinction between express methodological position and presupposed ontology, between theory and practice, and the latter demands explicit consideration too. That this is such a widespread feature of methodological criticism of Lawson, however, means that we must straightaway acknowledge that there is a deeper philosophical disagreement at issue in all these arguments, namely the acceptance or rejection of the form of reasoning that affords the examination of such ontological presuppositions: the transcendental argument (see Chapter 8).[9] In short, the argument presented here depends on an entirely different, *critical* conception of the role of philosophy itself, as opposed to either the 'first philosophy' conception of classical logical positivism or the antiphilosophical, scientistic or naturalising conception of post-Quinean philosophy.[10]

6.5 Conclusions

Let us recap. Lawson argues that, as for all sciences, a realist social science would be the examination of the real transfactual tendencies of the economy, and social reality is such that significant event regularities are not available and experimental closure is not possible. Mainstream economics, however, presupposes just such event regularities and so is fundamentally incompatible with its subject matter. For reasons of realism, therefore, in which our knowledge must fit to the reality it is knowledge of, mainstream economics is inadequate and needs reorienting.

Two conclusions follow that are relevant to the present project of an economics of science. First, as discussed above, the second refinement of our understanding of realism regarding the social sciences is that it must take a form that has ontological presuppositions compatible with the nature of social reality, which is constituted of social relations that are real but not actual. But in the case of the natural sciences, it seems that epistemic access to underlying causal mechanisms is attained by way of experiment, which affords reasoning to such mechanisms through the observation of artificially created event regularities. As such, a ('Hahnian') critic may still object that, even were Lawson's criticisms of mainstream economics to be accepted, they are idle for there is no way to learn about the social world other than to test theoretical conclusions against the actual course of events.[11] Given that social reality does not afford event regularities and the construction of event regularities in closed conditions is the primary means of access to underlying causal powers in the natural sciences, our second conclusion is that Lawson's ontological investigation of the nature of social reality raises the question: 'Is a realist social science possible?' In other words, how can a science proceed *other* than through observation of event regularities?

One major task for CR, therefore, is to spell out the implications of realism (i.e. ontological attention) for the methodology of the social sciences. Furthermore, CR is particularly beholden to this task because a central element of its argument for a transcendental-realist-inspired conception of science is that correlations are neither sufficient *nor necessary* for the identification of real causal powers (Bhaskar 1998). Lawson has defended himself against these objections by proposing a non-experimental methodology. We turn to this in the next chapter, where we argue that it is on this score that his argument leaves itself most open to criticism, *but for exactly the reasons identified in the CR critique of mainstream economics*. By way of immanent critique of Lawson's argument, therefore, it seems that the methodological implications of CR are that economics, as a social science, must be a *critical* project. Such a critical methodology is the final refinement to our understanding of realism as regards the social sciences and, in the process, decisively repudiates any recourse to deductivism, thus answering these 'Hahnian' objections.

Further reading

Fullbrook, E. (ed.) (2008) *Ontology and Economics: Tony Lawson and His Critics*, London: Routledge.

Kincaid, H. and D. Ross (eds) (2009) *Oxford Handbook of Philosophy of Economics*, Oxford: Oxford University Press.

Lawson, T. (1997) *Economics & Reality*, London and New York: Routledge.

Lawson, T. (2003) *Reorienting Economics*, London and New York: Routledge.

7

CRITICAL REALISM AND
BEYOND IN ECONOMICS

7.1 Introduction

In the last chapter, we considered the critical realist argument regarding the philosophy of economics and some of the major criticisms of that position. We saw that the unique focus of critical realism (CR) as ontological attention generates a novel ontology and epistemology that significantly challenges mainstream economics. However, the question remains regarding what is proposed methodologically in its stead. We shall argue in this chapter that, in considering this question, we must reappraise standard CR argument in ways that fundamentally qualify its conclusions but without entirely forsaking them. This argument involves two steps.

First, we shall reassess the positive suggestions of Tony Lawson (2003) regarding an alternative methodology for a critical realist 'economics as social theory', and argue, by way of immanent critique, that a critical methodology is stronger than the contrastive one he sets out. But what is meant by the terminology of 'critical' is then itself recontextualised in our second step, where we briefly reconsider the strongest aspect of CR philosophy, namely the derivation of transcendental realism from experimental scientific practice, in the light of constructivist studies of actual laboratory research practice from science and technology studies (STS).

This leads to a looser, messier and less conclusive grasp on the central philosophical issues discussed in this section. But in ways that are fundamentally in accordance with, rather than opposed to, the core tenets of critical realism regarding the limits of knowledge versus reality and an ontology of multiple interacting and constitutively relational (and so, insubstantial) emergent phenomena. The result is thus CR as a 'transcendental constructivism', in which the social sciences have the task of explaining the concrete socio-historical construction of given social phenomena and concepts, *including* those that are presupposed as necessary conditions of intelligibility of causally efficacious understandings of everyday social life.

7.2 Against contrastive methodology

Let us consider the existing methodological proposals of CR for a realist economics. As ever, given the priority of being over knowledge, such a methodology must allow a social science to be 'scientific' in the sense that it seeks to find the real underlying causes of social phenomena. In a bid to satisfy this criterion, Lawson has argued for the following 'contrastive' methodology: starting from the observation of partial or loose regularities (abbreviated 'demi-regs') that occasion surprise for the seasoned scientific observer (hence are 'contrastive'), the economist is invited to hypothesise an underlying tendency (i.e. to retroduce) that could possibly explain this unexpected empirical finding, before then testing it against competing hypothetical explanations employing the same contrastive methodology (Lawson 1997: Chapter 15, 2003: Chapter 4).

Lawson makes a number of important points in favour of this methodology. First, he is quite correct to point out the significance of contrasts as regards scientific explanation. For such contrasts not only (seem to) allow numerous non-experimental natural sciences to derive knowledge of their subject matter, thereby deflating to some extent the perceived problems associated with the lack of experimental social sciences, but also, at a more abstract level, are an important feature of explanation per se. As Lawson makes clear, since explanation is the rational reconstruction of the causal history of the phenomenon of interest, this, *in limine*, must lead back to the Big Bang. Given the absurdity of this demand, it is clear that an explanation is always asking for only the elements of the causal history that explain a particular aspect of interest, and this will typically be a matter of contrast: e.g. why X here while Y there? This, in turn, also points out the importance of rough or 'demi' regularities as points of departure in a social scientific explanation. Finally, the importance of 'interest' or 'surprise' in this account of explanation nicely highlights the centrality of pragmatic and/or practical context to CR epistemology that distinguishes it from a purely *ex ante* rationalism.

There are, however, various problems with the applicability of this conception to the social sciences, at least as the primary methodology. I focus here on the most significant of these, namely that it is not clear how the retroduced hypotheses are to be compared or 'tested' in any definitive or uncontroversial way. As discussed above, Lawson quite correctly points to the (generally accepted) success of numerous non-experimental natural sciences to show that hypothesised theories can nevertheless be tested against the manifest conditions of the particular subject matter in the world, i.e. outside the laboratory. He is thus perfectly justified to argue that the absence of social scientific experimental situations does not per se defeat the possibility of a social 'science' (Lawson 1997: 203–204, 2003: Chapter 4).[1] Furthermore, he correctly acknowledges that, given that the whole purpose of his argument is to *see* whether a social science is possible, he cannot start by examining any particular social scientific examination *presumed* to be an exemplary instance of successful use of his chosen model of science as contrastive explanation. But it is at this stage that he takes an illicit step in his argument because he thus chooses, 'for strategic reasons only' (Lawson 2003: 87), to illustrate contrastive explanation using the natural scientific example of plant-breeding.[2] But while his intentions are understandable, this decision is not as philosophically innocent as he suggests.

In order to see this, let us start with the theory-ladenness of observation and the related under-determination of theory by evidence. From the former it follows immediately, as Lawson notes, that scientific knowledge is produced in a dialectical learning process in which the redescription and production of evidence interacts with theoretical elaboration in an attempt to explain as much of the former as fully as possible using the latter (Lawson 2003: 101). Furthermore, he notes (1997: 215) that this 'dialectic' (or rather, messy parallel development) of theory and evidence immediately shows that the difference among the various sciences as regards the 'objectivity' of the evidence and the possibility of decisive test situations is not one of kind but of degree;[3] a point also brought out in terms of under-determination, which reminds us that it is perfectly *logically* defensible to object to any piece of evidence providing conclusive proof (whether falsifying or verifying) regarding a theory.

The under-determination of theory by evidence therefore requires some *other* (e.g. social) conditions, external to the judgement itself, to come into play if there is to be a (relatively) consensual closure regarding a knowledge claim, in order to rule out enough of the logically possible objections to make any such knowledge pragmatically viable. One such consideration is the relative autonomy of (judgements of) theory and evidence; the greater this autonomy, the more credible the reasoning from evidence to theory. But a realist must surely argue that these other factors are in the case of any given science (laboratory-based or not, 'natural' or 'social') also conditioned (at least in part) by the particular subject matter under study. It follows that there is no *ex ante* reason to presume that the relevant conditions for (the reality of) the *pragmatic* resolution of the logically *insoluble* problems of under-determination will take a similar form and/or be equally straightforward or arduous in each case. Rather, the intentional and value-laden nature of social reality suggests that the pragmatic resolution of problems of under-determination makes such an epistemology of testing hypotheses much more difficult for a social science than for the 'field trials' Lawson uses to illustrate his proposed contrastive methodology.

Indeed, if we acknowledge that the possibility of rational judgement is conditioned by how such judgemental rationality is itself conceived – what is epistemologically expected of a 'science' – then it also follows that an epistemology that demands of its knowledge the ability to achieve the logical impossibility of passing a definitive test situation may, in practice, pose a much lesser challenge for some sciences (where a pragmatically achieved consensus may be more easily taken for granted, and so overlooked as a necessary condition of a knowledge claim's uncontested establishment) than for others. Yet for the latter group, this imposition of such an epistemology will not merely make achievement of robust knowledge claims more difficult, but may frustrate it entirely by opening up any and every claim to a limitless wave of (logically impeccable) objections. In short, in the social sciences 'testing' of hypothetical theories, on the presumption that a conclusive judgement can be attained, may often be especially problematic, so that such a methodology may often make problems of *judgemental* (as opposed to epistemic) relativism particularly acute, and certainly qualitatively different to the field trials of Lawson's illustration.

Two issues regarding the nature of social reality are particularly important in this regard. First, in even the non-experimental natural sciences, the relative proximity and importance of purely physical evidence (which must itself, of course, be interpreted – e.g. what *counts* as a 'green' as opposed to a 'yellow' pea?) to the actual subject matter allows for a relatively easier establishment of the common grounds of understanding presupposed, not just by theoretical *consensus*, but also by an ongoing theoretical *debate* (with all that entails, including the various forms of social conditioning at play). Conversely, while never entirely isolated from the physical, the substantive concern of social science is with intentional phenomena, so that the role of theoretical interpretation in simply describing the 'facts' *before* they are then explained is that much more apparent and open to contestation. The result is not only that the use of 'stylised facts' is an irreducible element of a social scientific explanation but also that the question of *which* stylisation is acceptable is simply to take the debate *back* to theoretical argument, which it was hoped the 'facts' would resolve (see e.g. Section V, and especially Chapter 15 in Volume 2, for an illustration). Indeed, even the collection of data is subject to theoretical counterargument so that at no stage can rival theories be uncontroversially and definitively compared.

As I have stressed above, this is also a problem for the natural sciences but the intentional nature of the object of the social sciences can make the problems much more problematic; even, in some instances, effectively intractable insofar as the investigation claims to present a value-neutral 'scientific' analysis of the 'facts'. This is especially so given the (potentially) reflexive nature of social science, our second point, i.e. the interaction of social science and social reality and the possibility of the former, simply in describing social reality (whether 'accurately' and 'objectively' or not) to effect some change upon the latter. For this will often (but not always) raise more vividly than in the natural sciences issues of value (ethics, politics, etc.) that further complicate what will and will not be rationally conceded in theoretical argument. This close, if indirect, interaction of subject and object in the case of the social sciences (and thus between cognitive and practical, social issues) therefore indicates the essentially contested nature of both the conceptual starting places of social scientific investigation and the empirical record against which these are tested, not least because such investigations may have significant effect on socially-effective categories, which may be cherished or reviled.

For instance, consider the case of *The Spirit Level* (Wilkinson and Pickett 2010), a recent book that claims to document the negative social effects of greater social inequality and the benefits of its converse. This book received a triumphant reception in much of the liberal British media as providing a conclusive scientific case for the importance of reviving issues of social inequality in domestic policy and party politics. Moreover, the book employs a contrastive methodology similar to Lawson's prescriptions, making cross–country comparison of numerous metrics to explore the effects of social inequality, with apparently striking demi-regs regarding the correlations between greater social inequality and increased prevalence of other social ills.

Despite this exhaustive evidence and analysis, however, the report has – predictably – not established a new 'scientific' or 'evidence-based' agenda for British politics and the revitalisation of a politics of inequality but rather stimulated a series of critiques (including, but not exclusively, from right-wing think tanks and pundits; see e.g. Goldthorpe 2010, Hassan 2010, Kay 2009, Sanandaji *et al.* 2010, Saunders 2010, Snowdon 2010) contesting not just its policy implications, nor even the substantive findings, but the very methodologies and interpretation of the evidence, the 'demi-reg' contrasts motivating the study. This latter tactic, however, is both entirely predictable and legitimate, for Wilkinson and Pickett claim that the 'facts' they describe conclusively demonstrate the falsehood of deep-seated political convictions and understandings of the nature of society. Yet, both the intentional, reflexive and value-laden nature of social facts and the commensurate complexity and internal relationality of social phenomena mean that these understandings do not admit such definitive negation but are under-determined by any such findings. Whatever the merits of the case either way, therefore, it is certainly the case that *The Spirit Level* does not and cannot show, *beyond argument*, that greater social inequality is the only or main factor explaining all the other social problems discussed in that book.[4] The heated ensuing political debate about the book, therefore, is largely a symptom of the overstretched epistemology of the social sciences underlying both the report itself and that of many of its critics.

In Lawson's defence, he explicitly discusses these issues (e.g. Lawson 1997: 194 ff.) but in so doing he dismisses these problems by gesturing to the lack of a fundamental obstacle they have presented to the non-experimental natural sciences. This defence, however, does not work. Not only is it invalid to argue that because Peter faces the same problems as Paula, then Paula can cope with them as effectively as Peter, but the difference between these two cases is precisely the *realist* concern of their different circumstances and real natures. In the present case, what matters is the real nature of the subject matter of the non-laboratory natural sciences to which Lawson points and that of the social sciences. Moreover, Lawson's account of field trials gives insufficient attention to the pragmatic difficulties of achieving viable knowledge even in those non-laboratory natural sciences; problems that, were they given due attention, would suggest the even greater challenge they would pose for the social sciences.

In particular, the objects of the natural sciences are not themselves intentional and value-laden whereas the very categories of the *object* of study are always already political for the social sciences. This, in turn, makes scientific analysis of social phenomena in terms of given categories, as must be the case for a contrastive hypothetical method, much more problematic than for field trials, say, because the only way not to prejudge the validity of a given understanding of social phenomena is to interrogate the *construction* of these categories, their real social efficacy and the reality of their referents. Moreover, this also makes legitimate discrimination between 'factors' that must and need not be controlled for (setting the 'contrast space' in Lawson's (2003: 89 ff.) terms) when testing any such contrastive hypothesis extremely difficult, as any such judgement can always be challenged as presuming what needs to be proven; a problem exacerbated by the internal relationality of many social phenomena that Lawson notes, as all such factors must be externally related for this control to be possible.

Again, *The Spirit Level* perfectly illustrates this point: the bivariate analysis of inequality versus a plethora of other social variables (teenage pregnancy, health, alcohol abuse, etc.) deployed simply opens itself to the objection that the social reality is much more complex than such a model allows, especially at the level of whole nation states, and that the analysis simply 'proves' what is has already assumed, namely that inequality is the only relevant causal variable. The problem in this case, though, is not merely a matter of overblown claims or the absence of particular relevant factors in examination of specific correlations, but that such an objection is *always and in principle* possible and this therefore entirely undercuts the seeming 'realist' appeal of such a 'social science' as being able to provide the 'facts' to furnish *subsequent* political debate.

While we can concede that no science is entirely immune from problems of under-determination, therefore, it is perfectly defensible on critical realist grounds to argue that they will be *qualitatively* more pronounced in the social sciences. Indeed, even to the extent that we have acknowledged that *all* sciences, including paradigmatic natural, laboratory sciences, must grapple with under-determination, the foregoing argument simply allows us to turn this insight on its head and argue, with social studies of science, that natural sciences have many more of the problems usually consigned to the social sciences than is usually admitted, so that it is the context of natural scientific *debate* (in interaction with its subject matter) that makes these problems less apparent in most (but by no means all) natural science.[5] Conversely, Lawson's account of field trials appears to underestimate the problems, social and epistemological, associated with contrastive demi-regs even in the context of non-laboratory natural sciences.

In short, therefore, the problems of a judgemental relativism for the social sciences that arise from the nature of their object as itself intentional are simply not adequately addressed by Lawson's suggestions of hypothetical retroduction from contrastive demi-regs. There can be no expectation that such a method of testing hypotheses using contrastive demi-regs in the social sciences will lead to reasonably conclusive knowledge, even if it does thus succeed (not always, but with effort, argument, rhetoric and network-building) in various non-laboratory natural sciences. This is also evident in the fact that Lawson's contrastive methodology starts from social events that occasion surprise or curiosity, while in fact the pragmatic motivation for much social scientific research is not merely such cognitive concerns but to social phenomena experienced as *prima facie* unwelcome, if not normatively bad. Lawson's suggestions thus take too cognitive a stand, ignoring the crucial interaction of social *science* and social *reality*, and bracketing off issues of politics and ethics; issues, moreover, the *inclusion* of which was our original motivation to engage with 'realism' regarding economics. As we shall see, the only way to proceed, therefore, is to *admit* the political and international nature of social science from the outset and to incorporate these elements into our understanding of what a social 'science' is.

7.3 Refinement 3: For critical methodology

The irony of the foregoing criticisms, however, is that they are precisely Lawson's own criticisms when addressed to mainstream economics. For Lawson's contrastive methodology simply assumes and idealises a particular model of science taken from the non-experimental natural sciences and attempts to apply it to the social sciences, with the promise to deal, post hoc, with the effects of the disanalogies between the nature of their respective subject matters (cf. Ruccio 2005). Yet, this is precisely what the idealisation methodology of economics does, taking the *form* of scientific explanation it claims to observe in physics and attempting to apply it to subject matter for which it is quite inappropriate, rather than focusing on a methodology suitable to the task of scientific *retroduction* concerning that particular type of phenomenon. It follows that if Lawson was correct to criticise the latter position, as I have argued he is, he is also wrong to propose the former.

Where then does this leave the possibility of a realist social science? Lawson is correct that such a social science must (1) be concerned with the real, transfactual powers of social phenomena and (2) be compatible with the nature of social reality. Once again, therefore, we must attend to the nature of the subject matter for methodological guidance. But if we do that it is clear that the subject matter is always already interpreted, as discussed above, and this poses insuperable problems for a hypothetical/testing method such as that proposed by Lawson. The CR line of thought, counselling 'ontological attention', however, offers an alternative approach, because the very intentionality of the object of study entails that it also comes with ontological presuppositions and these can themselves be subjected to transcendental analysis.[6] This is therefore to propose a *critical* methodology on grounds of realism.

What is a critical social science? In answering this question, it is helpful so say what it is not: namely a *positive* social science aiming for definitive (if possibly fallibilist), general and neutral knowledge of particular social phenomena.[7] Clearly, such an ahistorical conception for social scientific knowledge has significant connections with a broadly positiv*ist* epistemology. But in what sense is the opposite 'critical'? First and foremost, such a social science is critical in the general Kantian sense of examining the necessary conditions of intelligibility of a given understanding and thereby *questioning* the 'surface' of such understanding, i.e. a *transcendental* enquiry. As discussed in greater detail in the Chapter 8, this remains fallibilist, however, because the given premises, upon which the conclusion depends, are themselves always contingent. It thus is also critical *of* an always-already given understanding instead of seeking to test self-generated first principles. As such, it proceeds methodologically by starting from the particular historical phenomenon of interest as described in our given practical understanding of it (hence imparting an irreducible hermeneutic moment to the analysis), and then seeking to uncover the various sociohistorically specific causes that explain why it occurred when it did and in the particular form that it did. This, of course, is a fundamentally constructivist move, inquiring into the historical construction of given concepts and social phenomena in their concrete empirical particularity. As a result, we may also call this critical methodology a 'transcendental constructivism'.

Furthermore, as a social science, insofar as it reveals underlying antagonisms and contradictions constitutive of that understanding it thereby punctures the bubble of any self-presentation of these social categories as harmonious, self-contained or 'natural'. It is also critical, therefore, in the sense that it may (contingently, and so may not) present a critique and negative judgement on that social arrangement itself, raising the practical syllogism which concludes in the necessity to change it in practice (though see Chapter 17 regarding the immediacy or swiftness of this conclusion). Finally, it is critical in the sense that it is most probably motivated by a practical problem and issues of theory/practice consistency. In this case, however, the (ideal of) theory/practice consistency works in both directions with our theoretical investigation stimulated by and measured against practice while changes in our practice are informed by our theoretical conclusions. I stress, however, that this critical value judgement is the *conclusion* of a realist social scientific investigation and not its *ex ante* premise or presumption (Bhaskar 1998: 7–8).

A critical social science thus includes as an element of its own methodology the examination of ontological presuppositions, so that the final refinement to our understanding of 'realism' as regards the *social* sciences in particular is this central role of ontological investigation in the substantive social science itself. A social science is only 'realist', then, on this conception where ontological analysis is involved to some extent in the genesis of some key theoretical categories themselves. But this is not to usurp social science with philosophy, for the development and elaboration of these concepts is the job of the social scientist in the ongoing dialectic of theoretical work and concrete historical detail, and this must take place at its own level of discourse, including the use of diverse methods for empirical research and triangulation.

The social scientific methodology that arises from a critical realist position, therefore, is once again to start with our *given* understandings and to subject them, inter alia, to transcendental argument (TA) in order to find out what their necessary conditions of intelligibility are and whether or not they are compatible with the surface understanding of our premise. Much, if not all, of this account so far is compatible with Lawson's account, especially as set out in Chapter 2 of *Reorienting Economics* (Lawson 2003), where he stresses the 'intelligibility' of the social world as the starting point for a critical realist economics. Here, however, we are not just employing transcendental reasoning in ontological examination of the presuppositions of (social) scientific methodologies and the construction of a (social) ontology (as does Lawson 2003: xvi) but *also* in a substantive way, working *from* understandings of significant social phenomena as are socially effective in those same phenomena, *to* the real necessary conditions of intelligibility of that understanding which constitutes the presupposed, relational social structure. As such, we attain knowledge of the real social structures that are the focus of a realist social science, identifying real, non-actual social relations that can then be the *starting point* for further constructivist, empirical research. Moreover, we do so in a way that not only addresses the (irrealist) problem of simply assuming and idealising our given understanding of social reality, but also thereby introduces a necessary discontinuity into the dialectic of theoretical and historical investigation that provides a relatively autonomous level of discourse to ground social science and so help to address to some extent, but never entirely, the otherwise intractable problems of judgemental (social) relativism discussed above.

There remains, therefore, the non-foundational interaction of historical and theoretical enquiry, but the latter is now tethered (however loosely) by the (social) ontological presuppositions of the self-understanding of its subject matter, actual social life. On this definition, then, it can also be seen how the terms 'critical' and 'realist' are complementary and mutually dependent. For it is only when a critical methodology is employed that a theoretical position may be deemed realist while, conversely, only when realism is preserved (and with it, judgemental rationality) can there be any force to a critical investigation. While this methodology is thus critical, and *so* realist, Lawson's contrastive, hypothetical methodology focuses on '*realist* social theory' (2003: 27, emphasis added) to the relative neglect of its *critical* aspect, or at least its limitation to a separate project of ontological under-labouring. A critical methodology thus incorporates and goes beyond Lawson's, in particular as regards the fundamental importance placed upon the already interpreted and valued nature of social reality and, inseparably, a *substantive* role for realist transcendental analysis. Contrastive hypothetical explanation is thus – like deductivist mathematical formalism before it – resituated as a useful but limited methodology that must be contextualised within a broader critical project.

In fairness to Lawson, contrast remains a crucial element of this methodology, but the contrast of principal relevance in most cases of scientific explanation will not be between different states of affairs but between different *theories* explaining more-or-less the same states of affairs.[8] But we have seen that even at this level, the social sciences are particularly beset by problems of judgemental relativism so that there can be no presumption that comparison of explanatory power (i.e. a purely epistemic criterion), while a necessary factor, will be able to resolve issues of theory choice. While comparison of explanatory power is crucial (e.g. Bhaskar 1998: 46), it is most unlikely to be decisive for the simple reason that its assessment in any one case is likely itself to remain controversial so that such comparison will tend to revert to discussion of *theoretical* problems in the various competing frameworks. The connection of the substantive transcendental argument thus remains vital, not least because the reversion to theoretical argument and comparison is thus broadened to include ontological argument, and this may afford more robust, if not conclusive, judgement given the element of (albeit contingent and conditional) necessity involved in a transcendental analysis and the fact that such arguments proceed, as immanent critique, from premises selected according to the critical context of ongoing debate (see Chapter 8).

Note also how this methodology necessarily incorporates a hermeneutic moment in which the social activity to be thus understood *must* first be understood according to the sociohistorically specific conceptualisation of participating agents, while it is not clear how such a moment is integrated into Lawson's methodology.[9] Against a purely hermeneutic social science, however, the transcendental derivation of what such understandings presuppose reveals social phenomena that must themselves exist '*intransitively* and may therefore exist independently of their appropriate conceptualisation and as such be subject to an unacknowledged possibility of historical transformation' (Bhaskar 1986: 51, original emphasis). Such social science is also, thereby, inevitably socio-historically specific, though this also follows from the contingent premises of the substantive transcendental argument.

Moreover, and no less importantly, such a critical methodology frames the interaction of any such investigation and the social reality it is studying very differently to the 'objective' study of a given social reality presupposed by the contrastive demi-reg methodology. In particular, the interaction of social science and social reality strongly suggests, against 'idealist' presumptions, that no amount of reasoned argument will lead to rational agreement on these issues (Bhaskar 1986, Fine 2010 on zombieconomics) and that social scientific knowledge is irreducibly conditioned by values and politics. This is insufficiently acknowledged by the contrastive demi-reg methodology but the critical methodology, by working from given understandings of social life and inquiring into their ongoing construction and transformation, self-consciously positions itself as participating in a never-ending, processual and intrinsically political process of social self-knowledge. The place for 'realist' social scientific knowledge, therefore, is to illuminate and participate in political debate but in the acknowledgement that such knowledge is itself an irreducibly political enterprise so that the possibility of defensibly rational and 'objective' judgement about social phenomena is actually opposed to, rather than demanding, (the pretence of) value neutrality.[10] Conversely, in positing the contrastive hypothetical method as the social scientific methodology that follows from critical realist ontological investigation, Lawson concedes too much to a positivist picture of objective and value-neutral scientific knowledge even as he effectively critiques it elsewhere.

In the present case of an *economic* social science, therefore, a realist economics would proceed from the examination of a particular historical problematic with a significant economic aspect (e.g. the changing economics of science) alongside the theoretical elaboration of the social structural presuppositions of the relevant economic form of life, in this case the commodity form. But the latter analysis already exists, namely Marx's *Capital*, hence:

> Marx's analysis in *Capital* illustrates the substantive use of a transcendental procedure. *Capital* may most plausibly be viewed as an attempt to establish what must be the case for the experiences grasped by the phenomenal form of capitalist life to be possible; setting out, as it were, a pure schema for the understanding of economic phenomena under capitalism, specifying the categories that must be employed in any concrete investigation.
>
> (Bhaskar 1986: 51)

In short, therefore, a critical realist approach to the economics of science licenses a relational Marxist methodology, which explores the ongoing construction of a socio-historically emergent and relatively autonomous 'economic' sphere characterised by capitalist relations of production that is simultaneously relational and structural; and thereby also providing a realist justification for the examination, in present circumstances, of the interaction of an emergent 'economy' and 'science'. While fuller justification of this conclusion remains to be provided (see Section V in Volume 2), we may note immediately that the critical methodology licensed by critical realism in economics has prima facie already been put into effect, substantially allaying doubt as to its possibility.

I want to stress, however, that this does not, of course, rule out other theoretical perspectives because Lawson (1999b: 14) is quite correct to argue that there is no one-to-one correspondence of philosophy (at least qua ontology) to social science, so that there can be no 'critical realist economics' per se. Indeed, to repeat, multiple (possibly innovative) methods are needed to develop empirical understanding of concrete actualities. Nevertheless, the theoretical 'big tent' to which recent CR pronouncements (e.g. Lawson 2003, Lewis 2004b) have appealed does seem overstated. In particular, it is (at best) an open question to what extent the critical methodology it is argued here that CR licenses is employed by all the various traditions to which Lawson appeals (e.g. Davidsen 2005), as evidenced, for instance, by the inadequate development, or even total absence, of the fundamental concept of (the social relation of) capital.

7.4 Some refinements in turn of critical realism – towards dialogue

The foregoing argument – for a 'transcendental constructivism' – also has some significant implications for CR as a project. Amongst the most important of these can be captured by considering an ambiguity in the central CR notion of ontology as 'underlabourer' to the social sciences. For it is possible to understand this either as the elaboration and articulation of a general social (and possibly economic) ontology that is then applied in its more or less completed form to questions of theory, or as a relatively autonomous but parallel philosophical discourse in which the (inescapable responsibility of) ontological understanding of basic (social) scientific terms, such as cause, tendency, mechanism, social relation, social structure, etc., are discussed.[11] No doubt, Lawson would see the role in the latter form, as is supported here. But there is an important corollary of this latter conception to which many of his pronouncements do not seem to give sufficient weight, namely that as a *critical* philosophical discourse, one must self-consciously expect that it will *never* be able to really pin down or perfect the understanding of these terms.

It follows that a truly critical parallel philosophical discourse must acknowledge that its primary role is not to 'underlabour' in the sense of *building up* the ontological foundations of theoretical work but, in a sense, the exact opposite: to 'undermine' or clear away internally inconsistent ontological understanding. Note also how this follows directly from assuming a realist stance in which reality is always greater than our conception of it, for this means that any and every conception of the nature of reality is *ultimately* fallacious and can only be relied upon to the extent that it is itself a *critical* point, i.e. a (relatively) determinate negation of our other always-already given beliefs (about the nature of reality), this then introducing an ineliminable dynamism to our ideas that prevents the fallacious (and irrealist) reification of any understanding as 'really' what the world is like.[12]

Against those who deem (social) ontology a pointless distraction from the 'real work' (e.g. Kemp 2005), however, this is not to deny the worth of the social ontological projects of Lawson and other critical realists but simply to recast it. These projects do not and cannot offer 'the' social ontology for our wholesale ingestion and application, nor even a 'separate' project of methodology/social ontology, but can and do present a *concerted critical reflection*, using critical and realist philosophical arguments, on what *is in fact* presupposed by the various social practices examined.

Certainly, it is the case that the leading protagonists of CR (such as Bhaskar, Lawson or Archer) would themselves subscribe to such conclusions. But there remains a tendency within such work to (seem to) offer positive and definitive pronouncements about the nature of reality per se. This tendency is, in turn, related to a second set of criticisms of CR, namely the lack of examples of those actually *doing* social science, of illustrations of the difference it makes to substantive social research and of evidence of the reciprocal change to CR through a *mutual* engagement with the social sciences. To an important extent, such criticisms are wide of the mark. On the one hand, it is unfair to demand of philosophers that they are also social scientists while also demanding that they answer questions and objections to their philosophical arguments (e.g. Hartwig 2009). Similarly, that ontological and substantive social scientific investigation occur in parallel licenses the argument that there can be no such thing as a single 'critical realist social science' that illustrates *the* difference ontological attention makes. Finally, there are also notable exceptions even amongst the leading critical realists, such as Jessop or Sayer, of those who are *primarily* engaged with a substantive research programme and so have explored the reciprocal impact of these on critical realist ontology. The number of social scientific projects that have explicitly employed critical realism is also impressive and growing.[13]

Yet, while there is indeed no such thing as a single critical realist social science, it remains the case that either CR ontological attention *does* make such a difference, however subtle or indirect, or it is pointless. Of course, even on Lawson's conception it does have a profound effect, fundamentally challenging the epistemic legitimacy not only of a mainstream discipline but of *the* most influential discipline of the social sciences. But on the argument of this chapter, that CR also licenses the use of substantive transcendental argument adds another dimension to the difference it makes, introducing a synthetic *a priori* conceptual/theoretical vocabulary needed for a realist social science.

Moreover, by paying explicit attention to the insights of constructivist, empirical social science on cognate issues, this not only addresses these two criticisms of critical realism head-on (namely the overstatement of definitive ontologies and the inadequate and one-way engagement with social science), but also revises critical realism in ways the make it open to more such engagement and thereby more attractive to researchers across the social sciences (including economics) who are not prima facie inclined to accept what may (rightly or not) seem to them as *ex cathedra* pronouncements of a grand ontological theory. While acknowledging the socio-political conditionality of knowledge, this may also make critical realism (even?) more effective in its reform of the social sciences than it has proven to date.

Focusing on the question of an economics of science is also a particularly productive arena for consideration of these issues, as the centrality to such a research programme of central issues for CR – such as the ontological questions of definition and (social) nature of science and the economy, or epistemological ones regarding the study of these phenomena – demands engagement with insights from the social sciences that have been empirically exploring such questions; an engagement we will illustrate in the following chapters. Before we proceed to these discussions, however, we can note three points that are both singular contributions of CR (to an economics of science), all arising from its signature call for ontological attention and the realist transcendental analysis it deploys, and positions that are themselves significantly resituated by such reciprocal interaction. These are:

1   The philosophy of science conclusions regarding ontological realism, together with transcendental realism (of a stratified and relational, hence insubstantial but emergent, reality) and (the possibility of) judgemental rationality despite (or rather because of, as enabled by) epistemic relativism.
2   A social ontology of emergent structures of socio-material relations.
3   A practical and engaged purpose or motivation to all such investigations, together with the forms of social critique afforded by the critical methodology advocated.

In each case, concerted and mutual engagement between such ontological investigation and insights from the empirical social sciences and the constructivist perspective they articulate and exemplify leads to (1) a welcome *deflation* of the CR conclusions but not one that *defeats* the CR arguments so that (2) the opposing viewpoints against which CR is sharpened must themselves take on board criticisms from CR. The result is thus a research programme of *mutual* lessons, albeit one that can have no expectation of conclusive movement towards an all-inclusive 'dialectical' synthesis. Indeed, on this conception, CR must not only critique premature analytical universality (as it does with great effect) but also resist the siren call of an equally premature (quasi-Hegelian) dialectical totalisation – a tendency that is arguably at work in the subsequent dialectical and spiritual turns of Bhaskar's work (1993, 2000, 2002). Instead, in the spirit of such deflation, CR is thus recast as a useful, indeed (contingently and conditionally) necessary but insufficient, conceptual resource – with all the pragmatic resituating this entails regarding a deflation of both the truth status of its conclusions and the ontological (including normative and political) reach of its arguments.

We will discuss points (2) and (3) in greater detail in subsequent chapters (in Volume 2). But let us briefly illustrate these arguments here by considering (1), regarding CR's philosophy of science, for this illustrates the way in which this engagement takes us beyond the critical realism exemplified by Lawson, with a reappraisal and reassertion of its *critical* aspect, towards the 'transcendental constructivism' discussed above. First, it must be (re-)stated that the philosophy of science schema originally articulated by Bhaskar (in *A Realist Theory of Science* (*RTS*)) is compelling and offers a determinate negation of the positivist and anti-metaphysical philosophy of science that dominated the twentieth century. It also, thereby, offers solid philosophical reasoning for the resolution of numerous philosophical problems regarding science, not least regarding the *possibility* of rational judgement in science despite all scientific work being irreducibly located in a particular social and historical context. This is a singularly significant contribution. However, the argument for transcendental realism works from a premise of a particular framing of experimental scientific practice. This premise was chosen, quite legitimately, in the critical context of particular debate in the philosophy of science in the 1970s. But subsequent, largely constructivist, analysis in the social studies of science has significantly challenged this picture, both as regards the practice of scientific experiment and the subsequent translation of scientific findings to the 'real world'.

First, the artificiality of the experimental situation is an issue that Bhaskar explicitly highlights as a crucial part of his argument; it is precisely because of the *need* for artificial experimentally closed conditions that he can proceed to argue that experiment makes manifest what would not otherwise take place and so reveals the non-actual real. Several seminal studies from STS, however, have shown that the artificiality of experimental set-ups is much more profound than this premise would suggest. For instance, Karin Knorr-Cetina (1981) argues that, far from nature being the object of study in an experiment, it is systematically excluded from the laboratory. Similarly Ian Hacking (1983) notes how even the paradigm case of physics works with phenomena in the lab that are almost entirely artificial. In the biological sciences too, the construction of experimental systems incorporates the very organisms under investigation, be it cell lines, micro-organisms or animal models that are specifically bred for experiments in highly regulated conditions.

The close interaction and dependence of experimental science on technological constructions also problematises the relationship between experiment and a mind-independent reality it is supposedly examining. Experiments increasingly involve laboratory equipment that has been developed purely and deliberately for experimental use, while the development of such technology has itself deployed prior scientific knowledge. Science, or rather 'techno-science' (Latour 1987, 1999), thus involves the interaction of parallel processes of construction – of technologies and of scientific knowledge claims. For some STS scholars, this even leads to a new form of the philosophical problem of underdetermination and the danger of a circular regress (e.g. Collins 1992), in which the only way a scientific theory can be tested is to rely upon technological apparatus, but the only way to test whether that apparatus 'works' is to see if it produces the results expected by the theory. While the extent to which this leads to an intractable philosophical problem may be challenged, it remains the case that experimental science is undeniably dependent upon prior human effort expended in the construction of highly artificial situations.

Indeed, another crucial element of the STS literature relevant to this question is the study of the history of the practice of experimental science and how it came to be accepted as the very paradigm of knowledge; for it was certainly not always thus. In the sixteenth and seventeenth centuries, when such practice was just beginning to become influential and to be theorised as uniquely important with the so-called 'Scientific Revolution', a common epistemological objection to experiment from the Aristotelian-inspired scholastic philosophy of the day was that knowledge about reality could only be ascertained by studying nature in its own setting. To attempt to isolate phenomena in the artificial context of a laboratory was thus to distort reality and so to produce 'knowledge' only of the artefactual phenomena thus generated. There is thus an important social history of how these objections came to be answered, or at least sidelined. Among the factors involved in such a history, however, are such contingent developments as the demonstration of experiments before select, socially elite audiences at the houses of English gentlemen, which lent the practice an air of authority and credibility (Shapin and Shaffer 1985). A further important element was the successful exploitation of knowledge generated in a laboratory setting in various 'real' world, e.g. military or commercial, settings. Indeed, the very idea of scientific objectivity has a social history (Daston 1992, Daston and Galison 2007).

This issue of the translation of scientific knowledge, itself another central plank of the Bhaskarian argument, however, has also been the subject of significant STS insights. In particular, numerous studies have explored how such application of scientific findings is dependent upon the entirely social phenomena of standardisation of technologies, organisational or institutional contexts and natural environments. For instance, Theodore Porter (1995) shows how the development of a science of forestry by the Prussian government in the nineteenth century depended upon the wholesale transformation of forest landscapes, replanting these in well-ordered rows; while Bruno Latour discusses how Brazilian ecological research depends upon efforts to turn the nature thus studied itself into a 'laboratory' or how assuring Pasteur's scientific success in debates regarding the biological sources of infection depended upon transformation of social practices within farms and the military (Latour 1999, 1993 respectively). In all of these cases, thus, contingent and highly socio-historically specific transformations in the social world are shown to be just as important as the experiment itself in establishing the authority of experimental results, both particular and in general.

Such studies significantly challenge the realist ontology and the epistemology of CR on at least two crucial points. First, they highlight the contingency of the construction of (all, including natural) scientific knowledge and its thoroughly social character, in interplay with the construction of a social order, in ways that go well beyond the simple invocation of the epistemic relativism of scientific knowledge. For this latter concept, connected with ontological realism and judgemental rationalism in a critical realist 'Trinity' (Archer *et al.* 1998 – with all the unwelcome religious overtones this brings), still leaves the distinct but misleading impression that the sociality of science is no impediment to establishing its *conclusive* epistemological justifiability. Conversely, when contingency and pragmatic social rationality are met at every turn in empirical accounts of actual scientific laboratory practice, this picture must be significantly revised.

Second, these studies significantly problematise the transcendental argument for transcendental realism, for if experimental science does not work as depicted in the philosophy of science discussion to which *RTS* is a contribution, the argument is robbed of its premise. Significant questions regarding the presumed unity of science from empirical evidence of the huge variation in 'scientific' practices (e.g. Dupré 1995, Galison and Stump 1996) add further to these questions.

The result of an engagement with empirical studies of science for CR would thus appear to be a move towards a Socratic deflation of its 'full-bodied' (Fleetwood 1999) (scientific) realism, contextualising it within a much more qualified and actively provisional acceptance of the truth of its philosophy of science. For instance, CR may be charged with overstating the conclusiveness of scientific (including social scientific) judgement and, relatedly, of closing down the (always) multiple understandings, at any time, of reality that are rationally defensible – offering too much philosophical consolation, as it were, regarding the possibility of closure of debate through rational argument and scientific work. Furthermore, while the philosophical constructivism of STS and its commitment to keeping explanation of scientific knowledge generation open has stimulated a whole field of insightful empirical inquiry, the CR philosophy of science too easily lends itself to the assumption that the problems of scientific knowledge have been fundamentally resolved, thereby prematurely foreclosing interest in, let alone actual undertaking of, such empirical research. An overstated realism, thus, may well be criticised for being an ideological block to research, rather than its precondition, as CR would seek to present itself (e.g. Sismondo 2007).

Such criticisms, however, do not entirely repudiate the insights and arguments of critical realism's philosophy of science. For the demand for ontological attention remains inexorable, with ontological presuppositions inescapable, even for the most radical of social constructivist accounts of science. Similarly, the transcendental argument for transcendental realism is merely deflated, rather than defeated altogether, for even given the artificiality of the closure conditions of experiment and the social constructivism involved in subsequently 'transfactual' applications of experimental scientific knowledge (both of which Bhaskar explicitly discusses, if with insufficient attention to their implications), it remains the case that reality is (relatively) amenable to description in the terms of such scientific knowledge (until, of course, there is sufficient social pressure in any particular case to tip the balance of judgement that the description is a 'failure'), and so we still need to be able to account for this fact that knowledge of phenomena generated in the artificial conditions of the lab can *possibly* be transferred to (the open system of) reality.

In fact, for all its insights into experimental practice, STS does not fundamentally challenge the picture of closure yielding results that require such closure precisely because they are non-actual. Rather, the challenge to the TR argument is the indirect one of undermining not the depiction of experimental practice that forms the premise but the *epistemic warrant* associated with this particular social practice, which makes it an appropriate target for transcendental analysis to ontological conclusions. To be sure, the latter is significantly qualified by the insights of STS, but this merely elicits a commensurate tentativeness in the conclusions rather than the complete abandonment of the argument. Furthermore, insofar as one is not totally incredulous of the capability of experimental science to elicit knowledge capable of application in extra-laboratory conditions – conditions, moreover, that may be greatly amenable to human manipulation, but are never entirely human constructions *ex nihilo* – one remains committed to transcendental realism, even while one may recognise its epistemic limitations. The conclusion of an engagement with STS on these issues is thus a position of provisional acceptance of the ontology and epistemology of critical realism, with a heightened awareness of the fallibility of these positions, rather than their abandonment in favour of, say, a radical ontology of neo-Kantian social constructivist idealism, as favoured by some (but by no means all) STS scholars. Indeed, the mutual accommodation of arguments demands of STS itself some significant changes to its philosophical pronouncements *towards* a CR position, the rejection or neglect of which has depended in no small part upon misunderstanding of the 'realism' of CR, as we discuss in the next and subsequent chapters.

7.5 Conclusion

We have argued over the past two chapters that a research programme (including an economics of science) that intends to yield scientific knowledge, in the cognitive sense of warranted belief about the contents of the objective world, must be self-consciously realist, and that this in turn depends upon examination of its ontological presuppositions to ensure that they afford such a mind-independent reality to be thus known. Given the real nature of their subject matter as always already interpreted, however, the methodological implications for the social sciences of the critical philosophy here advocated are that they must also be critical.

As regards the meaning of 'realism', then, from initial concern with ontological presuppositions as means of access to ontology, and hence preserving the privileging of being over knowledge, we have been led through a series of refinements to our understanding of the nature of the reality under study in a social science, and thus to our understanding of the nature of a 'realist' social science: a philosophical ontology of transcendental realism and constitutively relational emergent causal powers; a social ontology of transfactual social relations; and a methodology that incorporates a key role for substantive critical analysis. A 'realist social science' is one that satisfies all these requirements and so one that not only is avowedly realist, just as Lawson has argued, but also, going beyond Lawson's argument, one that employs a critical methodology, engendering a 'transcendental constructivism'.

Finally, in pursuing a concerted CR research program of ontological and philosophical immanent critique and mutual engagement with (theoretical) advances through empirical research in the social sciences – in the present case, those related to an economics of science – we have also been led to a significant resituating of these conclusions, transforming our understanding of the nature and significance of such a critical methodology; in particular, regarding the impossibility of theoretical or conceptual closure and hence conclusive critique. Indeed, by considering some preliminary concerns with core arguments of CR more generally we have seen how we must significantly revise, but not abandon, these arguments. In particular, engagement with constructivist studies of science have led to a significant resituating of the core CR argument from experimental scientific practice to the stratified ontology of transcendental realism. This move, however, has strengthened, rather than weakened, CR by making more apparent the non-foundational nature of its philosophical arguments, thereby substantially allaying not completely unfounded criticisms of CR regarding the extent to which it is genuinely a non-foundational research programme.

In this case, as with Lawson's contrastive methodology, however, the wrong turns we have discussed here reflect the omnipresent temptation to leap to premature cognitive closure, endemic to any realism (qua epistemic claim on the nature of reality), at the expense of the irreducibly *critical* approach to realism of 'critical realism' per se. Indeed, these constructive criticisms of CR serve to bring out a further sense in which 'critical realism' is 'critical', namely that a 'realism' remains epistemologically credible only so long as it is in process, and not declarative of established truths, hence critical *of* (a crude) realism.

In short, therefore, concreted engagement with a constructivist STS offers particular and *mutual* theoretical insights, which we shall continue to explore in future sections. One key remaining philosophical issue must be addressed, however, before we can proceed to these theoretical explorations, namely the status of the core philosophical innovation of CR, the realist transcendental argument. As we shall see, the relativisation of CR argument applies also as regards this central argument but, again, thereby paradoxically strengthening the validity of its reasoning.

Further reading

Bhaskar, R. (1998) *The Possibility of Naturalism* (3rd edition), London and New York: Routledge.

Fullbrook, E. (ed.) (2008) *Ontology and Economics: Tony Lawson and His Critics*, London: Routledge.

Lawson, T. (1997) *Economics & Reality*, London and New York: Routledge.

Lawson, T. (2003) *Reorienting Economics*, London and New York: Routledge.

Lewis, P. (ed.) (2004a) *Transforming Economics: Perspectives on the Critical Realist Project*, London: Routledge.

Sismondo, S. (2004) *An Introduction to Science and Technology Studies*, Oxford: Blackwell.

8

THE REALIST TRANSCENDENTAL ARGUMENT

8.1 Introduction

The philosophical debate at the heart of critical realism (CR) concerns the legitimacy of the mode of reasoning that it employs to reach its ontological conclusions: the transcendental argument (TA). This form of argument was most famously introduced to philosophy by Immanuel Kant in the *Critique of Pure Reason*, marking, as it were, the dawn of the modern era of philosophy in the (anti-) Copernican Revolution (Kant 1781/1953); just as Copernicus unseated 'man' from the centre of the universe with his heliocentric astronomy, so too Kant effected a similar blow to anthropocentrism by recentering the project of philosophy around the (self-)understanding of humanity and its world, rather than a metaphysics of reality per se.

The brilliant philosophical innovation of Kant was to pursue a completely different line of enquiry to the usual philosophical arguments of how different truths logically entail one another (or not) and instead investigate the *necessary conditions of possibility* of given premises. Traditionally, knowledge claims had been divided into those that are true by definition (analytic) and hence necessarily so or knowable in advance (a priori), on the one hand, and those that are informative (synthetic) but ascertainable only after empirical experience (a posteriori). Conversely, for Kant, transcendental reasoning produced an altogether novel kind of knowledge that was both necessary (a priori) and informative (synthetic). Transcendental reasoning and synthetic a priori truths are thus inseparable, the former licensing derivation of the latter. Kant argued that this in turn would resolve some of the intractable epistemological problems of philosophy, in particular the debates regarding the foundations of knowledge in reason (rationalism) or experience (empiricism).

But this august lineage by no means resolves the matter in CR's favour, and for at least two reasons. First, the status and success of Kant's philosophical project, including the validity of transcendental reasoning, remain seriously contested, especially by naturalistic philosophers of science, i.e. those committed to a wholesale deflation of philosophy as a *sui generis* epistemic project to its replacement with explanation of epistemic issues in terms of insights from the (social, psychological and/or natural) sciences. Second, CR's use of the TA in order to reach ontological conclusions about the nature of reality, i.e. a *realist* TA, is a significant departure from that to which Kant himself put it and to which he argued transcendental reasoning was uniquely suited (Parsons 1999a, 1999b, Viskovatoff 2002).

CR claims to use the Kantian 'method' of transcendental reasoning but apply it to distinctly non-Kantian premises and so reach equally non-Kantian conclusions. In particular, for reasons discussed below, Kant limited premises and conclusions to facts, supposedly self-evident to the thinker conducting the TA, about the individual mind of that thinker. Kant's conclusions thus similarly describe only features of the mind not reality per se, which remains entirely beyond human cognition. This therefore licensed a 'transcendental idealism', in which common sense realism about the reality of the objects of experience ('phenomena') could be preserved but at the cost of making the *conditions* for that realism ideal, or features of the mind; hence the 'ultimate' nature of things (or 'noumena') was beyond human conception. Conversely, CR claims to be able use the transcendental form of reasoning to examine the presuppositions of real (as manifest) social phenomena and thereby reach conclusions about the nature of reality, i.e. a transcendental realism. For CR, such arguments work to real conclusions about what must be the case *of/in the world* given contingently accepted a posteriori premises, hence yielding *conditionally necessary* ontological conclusions (viz. '*given* X, Y *must be* the case'). There is thus another debate regarding the validity of this transposition of the Kantian method to reach distinctively anti-Kantian conclusions.

In both cases, however, two issues are the main sources of contention: (1) the nature of the inference of necessity towards synthetic a priori conclusions – i.e. how does the TA work?; and (2) the ontological purchase of the argument – i.e. how does the argument make 'contact' with reality, as it were, and so give us knowledge about it? In tackling these profound and contested philosophical questions, let us consider one of most direct articulation of criticisms of CR's transcendental realist reasoning (Kaidesoja 2005, 2006). In doing so, I want to raise two ideas that are not discussed in the debate to date. First, I suggest a shift that we could programmatically call 'from Kant to Wittgenstein', and hence from discussing issues of epistemology to issues of meaning. Second, I want to suggest that a crucial factor underlying the different takes of naturalists and CR more generally is the different purposes to which philosophical inquiry is intended to be put.

8.2 Critical realism versus Kant

First, Kaidesoja reads (correctly I think, per Walsh (1975)) Kant as introducing transcendental reasoning as a form of argument that applies only to the singular question of 'what are the necessary conditions of possibility of experience?' for (at least) two reasons. First, transcendental reasoning is an examination of the structure of the understanding *by* the understanding itself. Its uniquely reflexive character thus limits it to examination of that which is given directly to the understanding, as it were, i.e. experience. Second, its transcendental (or aprioristic) character only arises in the context of an enquiry that yields apodeictically certain conclusions that cannot be unseated by any change in empirical experience: 'transcendental philosophy always proceeds a priori: its premises, justifications and conclusions are not based in experience. Transcendental philosophy consists rather in the a priori operations of pure reason' (Kaidesoja 2005: 35). As such, both the inference and, crucially, the premises of a TA must themselves be unquestionable, or else the conclusion can be refuted almost by the very act of querying the premise for this shows that, even if it ends up being true, it is not beyond question and the apparent necessity (a priori) of the conclusion collapses.

On Kaidesoja's reading, therefore, TAs are thus inevitably tied up with transcendental idealism, because a transcendental enquiry can only reveal how the mind itself of the transcendental enquirer must be structured for experience to be possible, i.e. what the mind must always already be like for it to be able to effect that particular enquiry. And since the mind is examining only itself, it can reveal truths only about the mind, yielding a transcendental idealism in which this limitation to conclusions about the mind rules out conclusive judgement about the nature of the reality that transcends the limits of the mind itself. But this immediately rules out, as incoherent, any TA deriving a transcendental realism: such a TA must necessarily overstep the limits of the mind in its self-understanding to make unwarranted conclusions about the nature of the noumenal reality. For Kaidesoja, it seems, although he appears to disagree with certain elements of Kant's analysis, Kant is essentially right about the nature of transcendental reasoning (if there is such a thing) and its limitation to examination of the conditions of possibility of experience and thus to transcendental idealism. He therefore charges Bhaskar and CR more generally with overstepping the bounds of transcendental reasoning – the attempt to separate the form of the TA from its particular Kantian content, it is argued, fails.

To this primary contention Kaidesoja also offers two further specific objections to CR's use of a realist TA. First, he argues that the nature of transcendental necessity apparently employed in the CR transcendental arguments is at best unclear and insufficiently articulated and defended, at worst simply Kantian transcendental necessity, in which case the realist TA does not work, as we have seen. He acknowledges that critical realists stress that the realist TA offers conclusions that are merely *conditionally* necessary, but objects that this defence does little to clarify the nature of the necessity at work in the realist TA. Indeed, he argues (2005: 38) that 'it is not at all clear how the notion of "conditionally necessary truth about the world" should be understood and how it differs from "fallible claim about the world"'. In which case, it is implied, a philosophical naturalism is really what is at work in CR.

Second, Kaidesoja accepts the Kantian position that experience is the only possible premise for a TA, or at least that premises must be self-evident or uncontroversial if the conclusion that X is a necessary condition of intelligibility of the premise is to follow. Accordingly, he argues that the description of scientific practice is, at best, not uncontroversial given the insights of science studies into the nature of scientific practice, including laboratory work that we considered in Chapter 7 (e.g. Collins 1992, Galison 1987, Knorr-Cetina 1981, Latour and Woolgar 1986), over the last few decades. It follows, he argues, that no a priori ontological conclusion can follow because the premise is not itself self-evident. Finally, Kaidesoja also argues that RB's use of immanent critique is irrelevant to the TA he offers. Instead, he argues, the TA consists only of the assertion that Y is condition of possibility of X.

In replying we can first immediately concede a number of points: that the argument that Kant takes the TA to be inseparable from transcendental idealism is quite correct; that CR's premise of scientific experimental praxis (in RTS) is not uncontroversial even if it is a faithful description of the self-representation of many scientists of their work; and also that the crux of the debate is whether the form of the TA can be deployed in non-Kantian contexts.

However, I want to suggest that there are two fundamental and connected problems in particular with Kaidesoja's position and interpretation of the TA and of CR alike. These are borne out of two crucial elements of the phenomenon of intelligibility, which is what is interrogated in a TA, that are systematically overlooked in mainstream philosophical accounts of transcendental philosophy, namely that intelligibility presupposes (1) a purpose or motivation and (2) a social location for the enquirer. Both of these elements contribute irreducibly to the critical context of a particular intelligible phenomenon, and this includes any transcendental enquiry itself.

8.3 Purpose or motivation in transcendental argument

Perhaps the most direct way to see the importance of explicit attention to the purpose or motivation in understanding particular perspectives or beliefs is to consider the false dilemma of first, foundationalist philosophy versus naturalist philosophy proffered by Kaidesoja and, indeed, most philosophers of science. For this dualism is only compelling so long as the shared motivation for engaging with philosophy that both of these perspectives assume is not brought into question. For mainstream philosophy of science it has now long been recognised (classically by Lukacs 1972) that its goal is a fully justified explanation of the rationality of science, which is taken as given and must simply be accounted for. Kaidesoja may object that naturalistic philosophy of science has no such goal, having completely repudiated any search for 'the' scientific method. Instead, naturalistic philosophy of science takes as its task the role of handmaiden to the sciences, assisting in their progress, including through the articulation of a coherent ontological theory. Yet, this latter task remains entirely beholden to the positive role of philosophy: to complete our knowledge in some respect.

Conversely, as I read CR, its primary philosophical purpose is a critical one. This involves accepting as given, sociohistorically specific and problematic the context of an ongoing debate and treating philosophy as simply one of the numerous epistemic discourses, albeit a particularly important one, with which we must engage if we are to apprehend and then (beneficially) transform our objective situation, whether personally or as a society. In each case, therefore, the motivation to turn to concerted philosophical reflection is the existing appreciation of a problem requiring resolution: e.g. for Bhaskar the inadequacies, aporia and even violence of positivism as a philosophical account of science; or for Lawson, the flagrant disregard of mainstream economics for the reality of its subject matter, and again, the normatively negative consequences of this. Against first philosophy, therefore, on this conception philosophy's status as *sui generis* does not license its total divorce and isolation from other questions, but rather the exact opposite is the case. It is because philosophy and philosophy alone can tackle questions that must be tackled in the course of *other* praxes (including, but not limited to, scientific ones) – i.e. that philosophy *is* a *sui generis* discourse but one that is important – that we must engage with philosophy. But its purpose and motivation lies without. Philosophy is thus primarily a *critical* enterprise (as indeed are all epistemic enterprises) and understanding a particular argument demands engagement with the particular circumstances that compelled the philosopher to embark upon their philosophical investigations.

Once this critical motivation is acknowledged, however, it becomes immediately apparent that there is a third option that is neither first philosophy nor naturalistic, but rather the examination of the philosophical presuppositions always already present in the ongoing and more concrete discourse at issue. Conversely, so long as the motivation of engaging with philosophy remains unquestionably the positive purpose of mainstream (and, not coincidentally, institutionally academic) philosophy, it remains effectively impossible that this third possibility can be entertained and understood.

Furthermore, it is only once the critical motivation of a CR philosophy is acknowledged that the nature of the conditional necessity of the TA can be accepted. For so long as it remains a matter of unspoken and evident acceptance that philosophy and the TA is aiming for certain positive knowledge, then 'necessity' will be interpreted as apodeictic and irrefutable and such a definition immediately makes unintelligible its conjunction with the qualifier 'conditional'. 'Conditional necessity' is thus not merely difficult to understand from this perspective but evident nonsense, a patent contradiction: the conditional unconditional. Conversely, once the critical purpose of philosophical investigation is acknowledged, the irreducible locating of the enquirer in a particular critical context becomes apparent and the goal of the investigation becomes analysis of what this particular self-understanding of a situation is presupposing must be the case in order to hold that up to question.

The goal is thus to challenge a given understanding not to perfect understanding once and for all; to reveal real, immanent contradictions in that given understanding not to show what really is the case.

The 'absolute necessity', which is taken as synonymous with simple 'necessity' by mainstream philosophy, is thus neither a motivating ideal nor the nature of the inference at work, making CR radically different to the concerns of mainstream critics. Yet, given the dependence of this point on the differing purpose of the philosophical enquiry in the first place, no amount of philosophical effort can make the latter perspective intelligible in the context of an unchanged commitment to a positive philosophy. And *this is something they cannot amend by more analysis* but only by a shift in position vis-à-vis philosophy that the philosopher must actually enact. The debate about TA in CR, their nature and relevance, is thus crucially conditioned by these differing purposes and there can be no expectation that argument alone will resolve the differences.

Another crucial illustration of the consequences of the different purposes accorded to philosophy is that they inevitably colour the interpretation of Kant. For the critical perspective, the singular insight of Kant was to offer a critical investigation, in which both the immanent critique and the transcendental argument elements of his investigation are crucial and inseparable. On this reading, Kant's argument, motivated by his twin concerns of 'the starry heavens above and the moral law within', is driven by the need to carve out the reality of a moral order in a seemingly determinist Newtonian universe, and thus to critique the philosophical positions that lead to this problematic, in particular the pretensions of 'pure' reason to be able to complete the human conquest of reality in knowledge and the defeatism of Humean scepticism. This he does by examining the necessary conditions of possibility of a premise given by the critical context of the philosophical debate to which he is contributing, namely experience: TA and immanent critique are thus inseparable. To be sure, *this* TA fails. But the crucial point is that Kant introduces a form of reasoning that undercuts the challenges, by asking of the enquirer what *they themselves* presuppose, and does so because he (i.e. Kant) himself is driven primarily *not* by the (ever-illusive) goal of apodeictic certainty but by the normatively intolerable conclusions of the contemporaneous philosophical debate.

Conversely, for the orthodox philosopher of science, the opposite is the case and it is the purely epistemological quandary that is primary in understanding Kant. As a result, the critical context of Kant's argument is not merely entirely overlooked but forcefully excluded – for the necessity for such a critical context immediately negates the apodeictic character of Kant's investigations. And this in turn means that it becomes impossible to interpret Kant as undertaking a critical enquiry but rather forces us to separate transcendental reasoning from immanent critique and thus focus entirely on the supposedly irrefutable nature of the former, which in turn pins it down to the single question of Kant.

Finally, as Kaidesoja exemplifies, the latter epistemological reading of Kant forces the TA to be understood purely in terms of the search for apodeictic certainty and thus limited to the transcendental idealism problematic of the necessary conditions of possibility of experience alone. Yet this does not work as Kant scholars have repeatedly shown (e.g. Brueckner 1996, Cassam 1987, Hintikka 1972, Stern 2003, Stroud 2000) and Kaidesoja himself concedes. He thus asserts the dualism of 'either Kantian TA or no TA' simultaneously with acknowledging the failure of the Kantian TA. The debate about the TA is thus already settled against it. Conversely, on the critical reading we can turn this dilemma on its head and argue rather that either the TA is *non-Kantian* or there is no such thing. This immediately opens up the debate about the TA beyond purely Kantian exegesis, and this takes us to the next point.

8.4 The sociality of intelligibility

We have seen, then, that the intelligibility of particular perspectives hinges on the purpose or motivation underlying it. But this leads immediately to the second point I would like to explore briefly, namely the irreducible sociality of intelligibility as a phenomenon. For with the irreducibility of purpose comes the irreducibility of the critical context of the investigation, and this will necessarily be sociohistorically specific. Indeed, a TA must take place in a particular language and built upon a particular understanding. This immediately shifts the debate from the early modern/Enlightenment, and excessively universalistic, terms of Kant so that it becomes clear that Kantian exegesis alone will not progress the debate. But once we recognise that examining the nature of the TA involves examination of the nature of meaning and language (and its relation to reality) we can move from critical discussion about Kant to critical discussion about Wittgenstein, from whom the insight that intelligibility is crucially intertwined with social and practical interaction, including with the natural world, arguably arises. In other words, I want to argue that critical engagement with Wittgenstein and the nature of intelligibility and understanding offers the possibility of acting as a bridge between the Kantian TA in the context of a subject-object dualism and a CR stratified philosophical ontology.

This shift would have to address two questions in particular, namely: (A) how does transcendental reasoning go from examining the necessary conditions of intelligibility to conclusions about the nature of reality, i.e. from intelligibility to ontology? And (B) how does the TA itself work? How can the TA be applied to other premises without forsaking its warrant given that, for Kant, it makes sense only when examining that which is given to the enquiring intelligence? I can only sketch my suggested answers to these questions here and no doubt numerous questions remain. But in reverse order:

(B) How does the TA work?

First, a brief recap. For Kant the TA works because it examines the necessary conditions of possibility of that which is self-evidently given to any intelligence possibly pursuing this line of questioning, namely the fact of experience. Since experience is itself partly constituted by the understanding, it follows that the understanding, which must be implicated in any attempt to understand something in particular, can come to knowledge of the necessary conditions of possibility of experience because it is a *self*-examination: experience lies within the grasp of understanding of the mind because the mind is itself involved in the formation of experience.

How then can the TA work when transposed to the novel context of the CR TA? In fact, for CR a similar logic emerges but built on the much broader understanding of the nature of the intelligence involved in such reasoning as an irreducibly social and practical phenomenon, and one in the process of development and growth. As such, the focus shifts from experience to intelligibility, because the conception of the nature of the enquiring intelligence is itself expanded. But this in turn entails a shift in the nature of that self which is subjected to and capable of transcendental reflection and *about* which we can reach conditionally necessary ontological conclusions: no longer is it the individual subjective experience but it is the embodied agent involved in intentional (and hence intelligible) discursive-material social practice.

As such, on this conception of the TA, the close and crucial connection between form and content necessary for the self-examining nature of the TA is not simply *severed*, which would defeat its epistemic legitimacy as a form of argument, but is *widened* in the realist TA. Both the relevant premise and the form of inference are expanded but still within limits, so that the integrity of the TA is preserved. There remain conditions of applicability of the TA, namely that the premise is a genuinely believed expression of a social practice that is acted upon, generally with successful fulfilment of intentions, as constitutive (at least in part) of that social practice. Social practices are always already conceptualised and so are real practices 'in the world' undertaken on the basis of a given understanding. Thus insofar as that practice exists in that particular form it is intelligible to a humanity capable of interrogating the *necessary conditions* of intelligibility of that practice in that form. It follows that insofar as the practice is genuinely performed it is accessible to examination of the necessary conditions of its *intelligibility*. Such an argument thus deploys neither logical (e.g. Quine 1980) nor analytical necessity (Strawson 1950, 1954, 1956, 1957, 1966), but what may be called 'intellective' necessity (Tyfield 2007), i.e. what must be the case given that a phenomenon X is intelligible *as* a phenomenon so that genuine belief that a given description of it is true underpins certain forms of social action. Thus focusing on the issue of intelligibility both preserves the reflexivity of Kant's insight into transcendental reasoning and elucidates it in a way that seems to offer a satisfying reflexivity.

(A) The purchase on reality

Turning next to the TA's purchase on reality, this has already been suggested in the foregoing and hinges on two points, namely that (1) the phenomenon subject to transcendental analysis on this expanded conception is a social, embodied practice that includes real interaction with the natural, mind-independent world and that (2) this is something capable of critical self-examination. It is the first of these, however, that is particularly important at this stage because it is the reality (in terms of causal power) of the ideational and its real interaction with the mind-independent world in social practice that licenses conclusions about the nature of reality, because reality is thereby not merely smuggled in at the conclusion but is there from the outset of the investigation, in the premise as it were.

To be sure, the reach of such ontological conclusions will depend entirely on the existing conceptualisation of the particular social practice at issue but this offers an important qualification and refinement of Bhaskar's (2008) seminal transcendental argument from scientific experimental practice; one, moreover, that responds to Kaidesoja's criticisms of this argument. For the reason that examination of the necessary conditions of intelligibility of *this practice in particular* yields a philosophical ontology of the nature of reality itself is because in the social context in which this argument is pursued, scientific experimental practice is *already* accorded the greatest epistemic status amongst all social practices. Insofar as the critical context of that argument is thus one in which scientific experiment is accepted as the paradigmatic social practice that informs us about the nature of the constituents of reality, it follows that the intelligibility of that practice yielding such objective knowledge necessitates a particular structure to the reality it is thereby elucidating. *Inter alia*, this entails that the philosophical ontology of stratified reality can only be expressed in the context of a society in which such scientific practice is *already* a powerful and successful social force so that the conclusion of transcendental realism is itself socio-historically specific and conditioned. But rather than undermining the ontological truth of CR's conclusions, this merely reinforces the plausibility of CR's express commitment to the conditionality and fallibility of its ontological judgements, just as we saw how situating this same argument in the light of the contingency and constructedness of science highlighted by science and technology studies (STS) did so in the last chapter.

8.5 Conclusions

In the light of the foregoing analysis, we can summarise the following features regarding CR's transcendental analysis. Proceeding from premises that express genuinely-believed and socially-efficacious conceptions of social practices, the realist TA examines what must (always already, hence a priori) be the case given the real intelligibility of these practices, thus formulated. The premises are thus always open to revision and debate, so that the conclusions of this ontological investigation are themselves always fallible and non-foundational. As such, to use the paradigmatic metaphor of a non-foundational epistemology, we remain aboard Neurath's boat of knowledge, which we must build even as we are sailing on it. By examining relations of necessity (i.e. what cannot *not* be the case if X is intelligible), however, the realist TA acts as something of an 'epistemic sea-anchor' that allows the intellect (itself embedded socially and materially in the world) to reach, however inexpertly, beyond the noumenal embargo on contact with 'reality' in the form of examination of ontological presuppositions, not least because it is, thus expanded through the sociality of intelligibility, always already embedded and embodied *within* (socio-natural) reality. By relatively anchoring philosophical (or social scientific) analysis in a different level of discourse, this therefore introduces an element of discontinuity in reasoning processes that can otherwise always be accused of being viciously circular (as we discussed in Chapter 7).

Furthermore, given that such presuppositions are inescapable, examining them in this way is simply to attend to beliefs, always already there, that have the potential to be the source of significant inconsistency between explicit position and implicit practical commitment. Transcendental realist philosophy is thus a necessary but insufficient step in the ongoing testing of such ontological presuppositions. Indeed, on this conception, the strength of the CR argument lies *not*, or at least not primarily, in the elaborate philosophical system of its transcendental realism and associated epistemology (viz. intransitive versus transitive dimensions of science, depth ontology, causal powers, transfactual tendencies, explanatory power, etc.), undoubtedly helpful though these may be. Rather, and in what is arguably an important qualification of 'orthodox' CR (e.g. Hartwig 2009 versus Datta *et al.* 2010), the strength of the position lies in its admission that the ontological understandings thus disclosed, even when refined through further argument, are *inadequate* (given that reality greatly exceeds knowledge) but also *inescapable*. Simply participating in a social world, including any form of intentional agency, is to commit oneself to an understanding of the nature of reality and one that is most probably (if not inevitably) wrong.

The conclusion (if it may be called that, given the epistemic dynamism and restlessness – even anxiety (Boltanski and Thévenot 2006) – it introduces, or rather acknowledges) of a critical realist philosophy is thus the diagnosis of an epistemic *predicament* that demands constant and ongoing work, given the normative obligations that arise from action based on (known or knowable) untruths, rather than construction of a definitive negation or critique. Such philosophical investigation works by orienting the investigator as much towards active engagement in the world as to one's responsibility for ongoing reflection. A critical realist philosophy thus demolishes not only any project towards the completion and perfection of knowledge, but also any hope of establishing conclusive and socially effective critiques *at the level of rational argument alone*. Moreover, it follows directly from the critical realist ontology – of mediated/ing, relational, insubstantial and non-self-subsistent emergent phenomena – that knowledge, *including* critical knowledge, is always imperfect.

This thus leads to one final point, regarding the sense in which CR may be said to be 'critical'. We have already considered several such senses, including its deployment of transcendental reasoning (including immanent critique) per Kant and its questioning, therefore, of the surface categories of everyday social experience. But against an interpretation of the latter as lending itself directly to a political programme of negative judgement upon social totalities, uncovered through this analysis to be major causes of social suffering, the above discussion suggests a further understanding of 'critical' social science that significantly qualifies such ambitions. For the conclusive status of any such critique is immediately significantly diluted by the broader conclusion of the greater distance at which knowledge claims are to be held. While not undermining such conclusions altogether, this deflation of the pretences not only of 'pure reason' or, conversely, empiricist knowledge but of thought itself attunes the critical realist much more acutely to the limits and fallibility of knowledge per se, including the conclusions of her own ontologically-derived critiques, so that these may now be interpreted only as informative but defeasible arguments, crucial but not definitive conceptual resources.

Such a conception of 'criticality' also resonates strongly with one that may be loosely associated with a lineage of 'genealogical' work that passes through, inter alia, Nietzsche, Foucault and Latour (see e.g. Vandenberghe 2002); namely to present the contingency of processes and events that have established the apparent 'necessity' of particular socio-technical arrangements. Such analyses are thus particularly attentive to both (1) the large and irreducible elements of contingency that mark processes of (historical) change and so stretch the explanatory capacity of analyses that seek to uncover totalities that, in turn, would license totalising critical judgement (in some cases) and (2) the fallibility and limits of knowledge and rationality (or rather knowledges and rationalities) per se, including as causal powers implicated in the phenomena under investigation.

This is not to suggest a wholesale embrace by a critical realist research programme of these 'genealogical' (pragmatist, post-structuralist, etc.) perspectives, for reasons we will explore in further detail in the following chapters. But we may conclude this discussion here by noting that analysis of the nature of the realist TA, the philosophical innovation at the heart of the research programme of a CR, itself leads to the suggestion that an engagement with these traditions may be a particularly fruitful line of future research, as we shall see in Section IV as we turn to explore the construction of an alternative, critical and explanatory economics of science and the crucial contributions of STS to this research agenda.

Further reading

Archer, M.S., R. Bhaskar, A. Collier, T. Lawson and A. Norrie (eds) (1998) *Critical Realism: Essential Readings*, London: Routledge.

Bhaskar, R. (1998) *The Possibility of Naturalism* (3rd edition), London and New York: Routledge.

Bhaskar, R. (2008) *A Realist Theory of Science* (2nd edition), London: Routledge.

Stern, R. (2003) *Transcendental Arguments: Problems and Prospects*, Oxford: Clarendon Press.

NOTES

1 Introduction

1 For instance, 'information economy', 'knowledge-based economy', 'knowledge-based bio-economy' (highlighting the particular emphasis on the life sciences characteristic of the discourse) and simply 'new economy'. These are themselves related to various terms denoting the break with the past that they represent, such as 'post-Fordist' or 'post-industrial' (Jessop 2005, Godin 2006).

2 See also subsequent chapters for more detailed discussion.

3 Among the now so-called 'BRICs' countries, the other two, China and Russia, were not involved in the WTO negotiations at that stage, hence their absence as further opposition to TRIPs.

4 As regards the social sciences, these changes thus augur the continual shrinking of spaces for intellectual inquiry as researchers find themselves upon 'academic treadmills' (Smith 2010) that channel their original motivating 'passion for social justice' into 'the pursuit of academic qualifications' (ibid.: 183, quoting Burawoy 2005: 260).

5 See, for example, Dasgupta and David (1987) David (1993, 1998), David and Dasgupta (1994) and Brock and Durlauf (1999), Diamond (1988), Wible (1998).

6 See, for example, Goldman (1986), Goldman and Shaked (1991), Kitcher (1993, 2000), Rescher (1989) and Zamorra Bonilla (1999).

7 This distinction is telling regarding the broader question of how there is now a discipline much of the mainstream of which has no interest in its ostensible subject matter, the economy; though we defer further comment on this point until Section III.

8 See, for example, Hands (1997, 2001), Fuller (1994), Kincaid (1997), Mirowski (1995, 1996), Mirowski and Sent (2002a, 2008), Sent (1997) and Solomon (1995).

9 This equivocation over 'information/knowledge' is deliberate, as it is a feature of this analysis not to distinguish between these (see below).

10 In fact, whether or not Bush should be credited with the linear model is now a matter of debate. See, for example, Dennis (2004).

11 'Positivism' is an unfortunate and imprecise term here, though I use it as useful shorthand. For those more familiar with the philosophy of science, this includes Popperian philosophy of science.

12 For contemporary defenders of Mertonian norm approaches, see, for example, van den Belt (2010).

13   For instance, perhaps you would want to object that a distinction could be drawn between information that is being exchanged and information that is being used for an act of exchange. Whether or not such a distinction is tenable – and/or tractable – is, however, doubtful.

14   See, for example, Etzkowitz *et al.* (2000), Etzkowitz and Peters (1991), Thursby and Thursby (2003) and Tijssen (2004). Gibbons *et al.* (1994) also arguably fit this description.

15   See, for example, Boyle (1996, 2003), Brown (2000), Campbell *et al.* (2002), Eisenberg (1987, 1996), Geiger (2004), Heller and Eisenberg (1998), Krimsky (2003), Lessig (2001, 2002), Merges and Nelson (1990), Nelson (2001, 2004), Newfield (2003), Orsenigo (1989), Rai (1999), Royal Society (2003), Soley (1995), Washburn (2005), Webster (1994) and references in Mirowski and Sent (2008).

16   See, for example, Dasgupta (2002) and Hahn (1992).

17   As we shall repeatedly see through this book, the simple dominance of a set of ideas is in no way testament to its intellectual coherence, let alone superiority (e.g. see Chapters 2 and 3 regarding the knowledge-based bio-economy (KBBE) and Chapter 4 regarding TRIPs). Similarly, the sociology of scientific knowledge (SSK) has also repeatedly demonstrated that there is no teleological guarantee that the winner of a scientific controversy will be the 'better' explanation (see Section IV in Volume 2).

18   Throughout this section, key terms in the diagram are in scare quotes to alert the reader to the caution needed to interpret the diagram and thus as a reminder to treat it only as a first step, not a definitive 'model'.

19   For example, the absence of discussion regarding *which* political economy and why (i.e. its epistemological justification) is the major lacuna in the otherwise excellent and insightful work of Mirowski and Sent on the economics of science, and thus a major contribution of this volume to the debate: see Sections IV and V in Volume 2.

20   For similar calls for analysis that is *both* relational *and* structural in some of these disciplines, see, for example, Bair (2005), Benton (2001) and Dicken *et al.* (2001).

21   The relation of presupposition is represented as the ovals within the various 'ideational' spheres and the vertical lines between explicit beliefs (the upper oval) and presuppositions (the lower oval). The box arrows denote the current and ongoing expansion and overlap of 'Economy' and 'Science'.

22   This argument parallels developments in critical realism more generally exploring the productive engagement with other 'post-structuralist' thinkers, such as Foucault, Derrida or Deleuze (e.g. Frauley 2007, Joseph and Roberts 2005, Sayer 2000).

2   The knowledge-based bio-economy

1   Moreover, both KBBE (discourse and reality) and the ensuing agrarian crisis are conditioned by a fundamental structural crisis of over-accumulation that has conditioned that emergence, and now crisis, of neoliberal financialisation.

2   For a more in-depth discussion of this argument and the particular 'transcendental constructivist' interpretation of the 'labour theory of value' adopted here, see Chapter 12.

3   On 'fictitious commodities', see Jessop (2007), Polanyi (1957).

4   Another important connotation is that value is socially necessary abstract *labour* time.

5   Relative surplus value refers to the increasing intensity of value production in a given period as opposed to absolute surplus value, which is rather the lengthening of working time. As the latter is quickly exhausted, it is the former that accounts for the majority of the explosive economic growth of (industrial) capitalism.

6   It follows also, against Brennan, that a machine or cyborg economy could not possibly be capitalist because such an economy would not create any surplus value. This, of course, is also a conclusion immanent to Marx's arguments about the law of the tendency of the rate of profit to fall with the tendential increase in the value composition of capital.

7 For critiques of other work in this literature, see Tyfield (2009) and Birch and Tyfield (forthcoming).
8 Indeed, once the completely different starting points of Brennan's and Marx's analyses are acknowledged, it can also be seen that Marx is not *himself*, arguing that 'nature' is dead or inert, but, like the argument regarding the nature/society distinction above, that it is necessarily *treated as such by capital*, not least because it is not 'productive' in the particular sense that matters from that perspective, i.e. productive *of surplus value*. In this way, Brennan's criticisms of Marx's anti-ecological bias are also quite wrong. For further discussion of the much-contested green credentials of Marx, see, for example, Bellamy Foster (2000), Benton (1996) and Burkett (1999).
9 It also serves to illustrate that, like 'nature' in the previous section, knowledge has always been crucial to economic activity, including under capitalism. This, of course, raises the question of what is thus special or new about the present 'knowledge-based' economy; a question we take up in the next chapter.
10 K = capital
    L = labour
    C = commodities
    Thick grey lines = main production output
    Thin dashed line = by-product of main production process.

    As in Chapter 1, this diagram must be interpreted as a first approximation of the role of knowledge in a capitalist economy, in order to stimulate further discussion, and certainly *not* as the basis for a definitive, ahistorical model.
11 The classic example of such a phenomenon is the open-source software movement, epitomised by the Linux operating system, Firefox web browser or Wikipedia. For a boosterist account, see Tapscott and Williams (2008); for a more a critical one, see Pedersen (2010) and Vercellone (2007).
12 This approach thus has clear resonances with the innovation studies literature on the importance of 'absorptive capacity' (Cohen and Levinthal 1990) and (the uneven, inequitable global) geography of innovation (e.g. Cooke *et al.* 2004, Moulaert and Sekia 2003).
13 Indeed, this also presents a singular problem for the very process of capitalist production of knowledge(s), including science, since the generalised commodity exchange of a capitalist market economy presupposes a real distinction between spheres of production and circulation.
14 As Jessop (2000b: 71) notes, 'the role of "intellectual technology" in the real subsumption of intellectual as well as manual labour' was already noted by Bell in one of the earliest discussions of knowledge-based (or rather 'post-industrial') economies (see Bell 1974: 29).
15 Note that, just as Callon (2002: 298) cautions regarding the relation between consolidated and emergent configurations, this is not to argue that there is some sort of natural 'cycle' or predictable progression between poles in any given case. Rather, the point is that the impossibility in the abstract of stable equilibrium imposes a constant restlessness to any capitalist knowledge-based activity, which rules out abstract prediction and demands its *actual* movement must always and only be studied in the concrete.

3 The KBBE reality

1 This is not to say that such sociological examination would be unwarranted, let alone impossible, (i.e. against the 'symmetry' thesis of STS – see Chapter 9) were we fundamentally to agree with its arguments, just that the need for such sociological enquiry is, pragmatically, obviously all the more pressing where we do not.
2 One result of this change was that, in 1995, such was the economic importance of finance, *Fortune* magazine decided to include financial companies in its list of the top 500 US companies (see Davis *et al.* 2003).

3 This trend continued up to the Great Crash of 2008. Since being forced to sell off their financial arms in the last few years (which are themselves now in severe trouble following the collapse of the American credit market), American car firms have continued to suffer catastrophic losses, leading the extraordinary bail out of GM by the US federal government (*The Economist* 2009a).

4 It is possible that apologists may try to argue that finance is itself knowledge work, though this would be to stretch the credibility of the argument. As discussed in more detail below, however, there are crucial links between the KBBE and finance.

5 For instance, is the outsourced cleaner at an NHS hospital a private or public-sector worker? What about the banker at the nationalised Royal Bank of Scotland? Or the university librarian, where universities receive a considerable chunk of income from both the state (central or local) and from 'private' individuals (student fees, commercial or charitable research funding, consultancy fees, etc.)?

6 These observations are particularly germane regarding projects in the Economics of Science within the Austrian tradition, following the work of Hayek (1937, 1945). Such studies seek to turn criticisms of the commercialisation of science on their head by showing that public, i.e. government, sponsorship of science invests too much power to set scientific agendas in a single authority and so distorts scientific advance, while a free market of ideas, with multiple, much smaller (than the state) private funders, approximates much better to the optimal allocation of financial resources, including from the perspective of epistemic virtues of the resulting science. While perfectly justified in the argument that public funding of science is political, such studies completely overlook: (1) the concentration of economic power in the private sector as opposed to their fantastical model of free competition; (2) the resulting complicity of 'state' and 'market', so that no such strict distinction is tenable, robbing this analysis of its premise; and (3) that the present problem regarding science funding is not that science is being *made* political, but the limitation of political agendas for science to those held by private (legal) persons with concentrations of financial wealth.

7 It is important to note, therefore, that while some Austrian economists, such as von Mises, argued only for the *reduction* of the role of the state in the economy, others, including Hayek, argued for a *different* role.

8 On discourse and the construction of a 'new' capitalism, see also Boltanski and Chiapello (2005) and Chouliaraki and Fairclough (1999).

9 This relates to Keynes' comparison of the stock market to a beauty contest in which the challenge is not to choose your own favourite but to guess the most popular choices of everybody else.

10 Although only 15 per cent of global agricultural produce actually is traded across borders, the size of global markets is such now that they affect the prices for almost all of the remaining 85 per cent as well (Directorate-General for Agriculture and Rural Development 2006).

11 A notable exception is vitamin A-enhanced ('Golden') rice.

12 SNLT = socially necessary labour time of the production process, as determined by market competition.

13 Another classic example of this process would be the IPCC reports on global climate change and the 'Climategate' scandal at University of East Anglia.

14 As Barry and Slater (2002: 185) note, measurement and calculation are often intended to have anti-political effects and often succeed, but cannot be guaranteed to do so since the 'framing is always in principle contestable'.

15 For example, one can imagine a predictable response of KBBE discourse to the current 'fiscalisation' of politics in the EU in aftermath of 2008 great crash – where reduction of structural fiscal deficits has become the utter priority of all government policy – as a new and compelling argument for even more radical dismantling of public research infrastructure towards the construction of a fully privatised KBBE as saviour of a new 'Green New Deal'.

4 Intellectual property rights and the global commodification of knowledge

1 An earlier version of this chapter was first published as Tyfield (2008a).
2 For example, Maskus (2000).
3 Given the importance of the life sciences, therefore, another industry that is of huge importance regarding the globalised commodification of knowledge is the agri-food business, as described in Chapter 3. It was the pharmaceutical industry, however, that was particularly instrumental in the enactment of the TRIPs agreement, hence the focus of this chapter on so-called 'red' (i.e. medical) rather than 'green' (i.e. agricultural) biotech.
4 Sell (2003) provides an excellent summary. See also Drahos and Braithwaite (2002), Dutfield (2003), Matthews (2002), May (2000) and Richards (2004).
5 Reddy (2000), following Hymer (1975, 1979). The resonance between globalisation and knowledge-intensive activities, including science itself, has been widely observed: e.g. Carnoy (1993a, 1993b), Castells (1993, 1996, 1997, 1998), Dicken (2007) and Drori et al. (2003).
6 For instance, as early as 1957, Pfizer already had overseas sales exceeding US$60 million (Drahos and Braithwaite, 2002: 66).
7 The off-shoring of R&D centres of TNCs to developing countries, especially China and India, has increased significantly in the last few years (especially from around 2007) (The Economist 2010b). However, such 'globalisation of innovation' remains an embryonic development and was certainly a negligible phenomenon even to the end of the twentieth century (Patel and Pavitt 1998).
8 It is extremely difficult to be more precise than this because R&D figures are not readily analysable from the filed accounts of the firms (see Angell 2005: Chapter 3). Furthermore, a high-profile estimate of current drug development costs by DiMasi et al. (2003, 2005a, 2005b) at $403 million (or $802 million if capitalised) is controversial for reasons of methodology; see Angell (2005: 41), Light and Warburton (2005a, 2005b) and Public Citizen (2001b).
9 See, for example, Taylor and Silberstom (1973), quoted in MacDonald (2002: 23); and Cohen and Merrill (2003), Cohen et al. (2002), Levin et al. (1987) and Mansfield (1986).
10 See, for example, Angell (2005: xxiv), who notes the importance of another form of monopoly rights provided in the US to address this problem, namely the exclusive marketing rights from the Food and Drug Administration (FDA).
11 Sell (2003) makes a similar point.
12 See Gallini (2002: 146), Kenney (1986: 257) and Orsenigo (1989: 46).
13 In fact, such patenting was not completely prohibited, but was allowed only after a laborious administrative process in which special approval was granted to patent (Slaughter and Rhoades 2002: 85).
14 On fears of Reagan's cuts, see Kenney (1986: 28).
15 The percentage of basic research R&D funding going to the life sciences rose in the 1970s from 36 per cent to 44 per cent – in parallel to the increase in the percentage funded by the NIH, from 36.7 per cent in 1971 to 47 per cent in 1981 (Mowery et al., 2004, using National Science Board data) – while that going to physics fell from 18 per cent to 14 per cent: Mirowski and Sent (2002a: 24). 'NIH' is the National Institutes of Health, the primary federal funding agency for the life sciences in the US.
16 See Calvert (2004) on the basic/applied science distinction.
17 For example, Mowery et al. (2001), esteemed economists of innovation and no political radicals, argue that these reforms were based on 'a belief by policymakers (based on little or no evidence) that stronger protection for the results of publicly funded R&D would accelerate their commercialisation'. See also Eisenberg (1996). As such, the relevant question becomes how this entirely outdated, if not disproven, argument became the seemingly authoritative argument for singularly far-reaching policy changes.

18 See, for example, Heller and Eisenberg (1998), Mazzoleni and Nelson (1998), Nelson (2001), Orsenigo (1989: 84). Cf. also discussion of some of the contradictions of the capitalist KBE in Chapter 3.

19 See, for example, Cohen et al. (2002), Klevorick et al. (1995), Levin et al. (1987), Rosenberg and Nelson (1994).

20 Note how this point arguably undermines the argument of Perez (2002), regarding new techno-economic paradigms that are 'waiting in the wings' (see Chapter 15).

21 Data from PhRMA (2005).

22 Compare DiMasi et al. (2003, 2005a, 2005b) with Angell (2005) and Light and Warburton (2005a, 2005b).

23 Data from Fortune 500, author's calculation.

24 For comparison, in 2001, 35 per cent of PhRMA revenues were spent on 'marketing and administration', of which roughly three quarters was marketing (Angell 2005: 120, using PhRMA data), hence 27 per cent. Conversely, R&D represented 16.7 per cent of total revenues that year (PhRMA 2005).

25 Examples of minor modifications remarketed under new brand names include Clarinex to Claritin, Prozac to Sarafem, Prilosec to Nexium (Angell 2005: 76 ff.).

26 For 'a damning case, not just against the industry but against our [US] entire system for developing, testing and using prescription drugs' (Angell 2006) see, for example, Abramson (2004), Avorn (2004), Goozner (2004), Kassirer (2004), Moynihan and Cassels (2005) and Olfman (2006).

27 Bud (1998: 14) pins down the start of this financial frenzy exactly to June 1979 when Nelson Schneider of investment house E F Hutton heard about Genentech's production of human insulin, became interested and reported biotech to investors as a major technological breakthrough.

28 For a full discussion of the case and the effect of the judgment, see Krimsky (2003: 62 ff.).

29 This is not to assert that legislative changes caused such a surge. For arguments against this interpretation, see Mowery et al. (2001, 2004) and Kortum and Lerner (1999). Rather from the perspective presented here, the argument is that the Bayh-Dole Act must itself be explained as one element of the cultural political economic construction of the neoliberal mode of regulation for the primitive accumulation of (biotech) knowledge.

30 Jaffe (2000: 543) and author's calculations using USPTO data. Note also that this is just for one class of biotechnology patents.

31 Rai and Eisenberg (2003: 300) report that, in 2000, the top five universities grossed nearly one half of total licensing revenues, while Thursby and Thursby (2003) note that, in the same year, only 43 per cent of licences earned royalties at all, and only 0.56 per cent earned over $1 million.

32 See also Mowery et al. (2001: 104) and Owen-Smith and Powell (2003: 1697), who note that biotech patents were 49.5 per cent of all university patents.

33 Federal funding of total US R&D fell below the 50 per cent mark in 1979 and continued to decline to a low of 25 per cent in 2000, while funding from private industry has taken the opposite path; see NSF (2006).

34 For instance, Angell (2005: 71) notes that the universities are just as resistant as big pharma to rigorous enforcement of a clause in the Bayh-Dole Act that demands 'reasonable terms' for the contracts resulting from such patents, as this may offend their big pharma sponsors.

35 At approximately 50 per cent of investment in Europe in 1999 as opposed to 'basic' at 12 per cent, cell factory and plant biotech each 9 per cent and animal biotech 8 per cent: Senker (2000: 57). Dutfield (2003: 146) reports 60 per cent of US and EU biotech firms produce health-related products.

36 See also Blackburn (2006) and Harvey (1982, 2003).

37 For discussion of 'historic bloc' and 'primitive accumulation', see Bieler and Morton (2004), Cox (1987, 1996), Gill (1993a, 1993b), Gramsci (1971), Harvey (2003), Jessop (1997) and Jessop and Sum (2006).

38 Sell (2003) discusses a similar phenomenon as regards US trade legislation at the time, which during the TRIPs negotiations continually upgraded the privileged position US trade policy provided big pharma. Kenney (1986: 242) also notes the irreducible role of the state in the development of biotech. See also the discussion of Mirowski and Sent (2008) regarding a tripartite analysis of relations between academia, industry and government.

39 This is not to deny the importance of EU or Japanese pharmaceutical companies regarding the actual TRIPs negotiations, and their own respective domestic political agency. Nevertheless, the TRIPs agreement was overwhelmingly an American initiative (Sell 2003) and I focus on the US aspect for lack of space. For an excellent comparison of the US, UK and Germany regarding biotech and the commercialisation of the university, see Jasanoff (2005). Nor is this to argue that the economy can be simplistically understood in terms of a fixed and homogeneous dualism between 'core' and 'periphery' (see e.g. Arrighi and Silver 1999).

5 Privatising Chinese science

1 Compare the important slogan throughout China's modern history of 'Chinese essence, Western means' (*zhongyi xitong*).

2 Particular difficulties associated with studying Chinese science include: the rapid pace of change; a system that is not easily amenable to standard Western categories of social theory; unreliable, and a general lack of, statistics; sheer size and regional diversity; and impenetrable political decision-making processes.

3 This political structure has been characterised as 'fragmented authoritarianism' (Lieberthal, 1992) and 'local state corporatism' (Oi, 1992).

4 This is also true of 'non-governmental technology enterprises' (*minying keji qiye*) (Segal 2003), which have been the backbone of much of China's technology reforms.

5 Note that the comparative methodology of this analysis follows both the critical realist arguments of Tony Lawson regarding a comparative, explanatory economics and Sheila Jasanoff regarding comparative studies of science policy (see Sections III and IV in Volume 2 respectively).

6 China has already approved GM cotton, tomato, petunia, papaya, poplar and sweet pepper, but the key crop of rice in particular has been subject to these significant delays. Insect-resistant GM rice was issued with biosafety certificates in November 2009. However, the official ambiguity continues as discomfort with the decision at the 2010 National People's Congress forced a senior rural affairs officer to make a statement reassuring the public that GM rice was still 'a long distance away' (*China Daily* 2010).

6 Towards a critical realist economics

1 We may note immediately that particular sense attached to 'ontology' and 'ontological commitment' by CR; as 'the understanding of the nature of being per se that is presupposed' *not* 'the type of things that exist to which my beliefs commit me'. Given that the latter connotation of ontological commitment has a long history in the (Anglophone, (post)-positivist) philosophy of science (particularly following Quine 1980), I will use 'ontological presupposition' instead in this chapter.

2 Conversely, this form of argument also, in some circumstances, licenses the conclusion that the presupposed phenomena must themselves exist, given the intelligibility of the premises – hence the presence and importance of presuppositions in the schematic diagram discussed in the Introduction (Figure 1.6).

3 The term 'transcendental realism' thus captures both its method of inference or derivation, by transcendental argument, and the fact that it ascribes the belief in a reality *beyond* the empirical. For reasons that cannot be discussed here, it is not, therefore, the 'transcendental realism' criticised by Kant.

4 Note, therefore, that this also includes *counterfactual* things, states of affairs and events, while the 'actual' and 'counterfactual' may be conventionally understood as opposites. In this way, critical realism is distinctive from the 'realism' of Mill, Cartwright or Hausman.

5 For a fuller discussion of these issues, see Bhaskar (1998, 2008) and Lawson (1997).

6 The critical realist literature on economics is large and growing. See, in particular, Lawson (1997, 1999a, 1999b, 1999c, 2003) and the collections of Brown *et al.* (2002), Downward (2004), Fleetwood (1999), Fullbrook (2008) and Lewis (2004a).

7 Lawson (2003: 7–11) offers quotations from, inter alia, Ariel Rubinstein, Wassily Leontief, Mark Blaug, Ronald Coase and even Milton Friedman to this effect.

8 See also Hands (2001: 323) for discussion of similar criticisms from others, and Vromen (2004).

9 For instance, Vromen (2004) expresses a common criticism when he asks for a defence of the transcendental arguments Lawson employs.

10 See, for example, Cruikshank (2002) for further discussion of these distinctions, with particular regard to the critical nature of a critical realist philosophy. Note that in the absence of the transcendental argument, philosophers tend to see in a critical philosophy only the opposing form of philosophy to their own: first philosophers see a naturalistic philosophy and vice versa.

11 For example, Lawson (1997: 18) quotes Hahn to this effect. See also Lawson (2003: 83).

7 Critical realism and beyond in economics

1 See also Benton (1981) for a critical realist argument for the importance of not overlooking non-laboratory natural sciences in construction of an epistemology.

2 Another favoured example is the contrastive explanation surrounding BSE, 'mad cow disease'.

3 This point is also fundamental if a critical realist philosophy of (social) science is to do more than pay lip service (through the concept of 'epistemic relativism') to the social conditioning of the production of scientific knowledge. For it is only in the light of this admission that it can incorporate the various developments in science studies that show how even paradigmatic cases of (natural) scientific experimentation produce knowledge in ways that do not allow the evidence simply to 'speak for itself'. For a fuller consideration of the social studies of science, see Section IV.

4 I would add, as an aside, that I am entirely sympathetic with the overall thrust of the argument in *The Spirit Level* regarding the politically overlooked importance of inequality; but this does not change the validity of the methodological criticisms that have been rightly leveled against it.

5 Consider, for instance, debates about climate change. This argument resonates strongly, of course, with the argument for treatment of 'natural' phenomena as 'socio-nature', discussed in Chapter 2.

6 The present argument thus goes beyond Bhaskar's (1998) argument that the already-interpreted nature of social reality affords a head start to the social sciences that compensates to some degree for the absence of experiments.

7 Such a conception, therefore, has a significant similarity to the critical versus problem-solving distinction regarding theory in international relations or international political economy posited by the neo-Gramscian scholar, Robert Cox (1996: 88–89).

8 This is not deny that contrastive demi-regs regarding states of affairs may not also be highly informative, but simply that they are unlikely to be able to play the role of definitive hypothesis testing suggested by Lawson.

9 I note that Lawson (1997: 34–35) expressly acknowledges that a realist economics must be sociohistorically located, including a hermeneutic moment. The problem is that his methodology does not in fact incorporate these points.

10 For more discussion on the interaction of social science and politics, see Volume 2, especially Chapter 17.

11 For example, the special issue of *Journal of Critical Realism* (2009) on causal powers.
12 Again, I note that Lawson (1997: 211–212) expressly aligns himself with such a critical conception of knowledge.
13 For example, Carter and New (2004), Joseph (2002), Patomäki (2002) and Sell (2003). See also Lawson (2003: 302, note 39) for a list of references exploring the redefinition of key economic categories in critical realist terms.

REFERENCES

Abramson, J. (2004) *Overdosed America: The Broken Promise of American Medicine*, New York: HarperCollins.

Adviesrad voor het Wetenschaps- en Technologiebeleid (AWT) (2005) *Een Vermogen Betalen: De Financiering van Universitair Onderzoek*, Den Haag: AWT.

Albo, G. (2007) 'The Limits of Eco-Localism: Scale, Strategy, Socialism', *Socialist Register* 43: 337–363.

Altenburg, T., H. Schmitz and A. Stamm (2008) 'Breakthrough? China's and India's Transition from Production to Innovation', *World Development* 36(2): 325–344.

Angell, M. (2005) *The Truth About the Drug Companies: How They Deceive Us and What to Do About It*, New York: Random House Trade Paperbacks.

Angell, M. (2006) 'Your Dangerous Drugstore', *New York Review of Books*, 8 June 2006, 53(10).

Archer, M.S., R. Bhaskar, A. Collier, T. Lawson and A. Norrie (eds) (1998) *Critical Realism: Essential Readings*, London: Routledge.

Archibugi, D. and S. Iammarino (2002) 'The Globalization of Technological Innovation: Definition and Evidence', *Review of International Political Economy*, 9(1): 98–122.

Archibugi, D. and B.-Å. Lundvall (eds) (2002) *The Globalizing Learning Economy*, Oxford: Oxford University Press.

Arrighi, G. (1994) *The Long Twentieth Century*, London: Verso.

Arrighi, G. (2003) 'The Social and Political Economy of Global Turbulence', *New Left Review* 20: 5–71.

Arrighi, G. (2005a) 'Hegemony Unravelling – 1', *New Left Review* 32: 23–82.

Arrighi, G. (2005b) 'Hegemony Unravelling – 2' *New Left Review* 33: 83–116.

Arrighi, G. (2008) *Adam Smith in Beijing*, London: Verso.

Arrighi, G. and B. Silver (eds) (1999) *Chaos and Governance in the Modern World System*, Minneapolis, MN and London: University of Minnesota Press.

Arrow, K. (1962a) 'Economic Welfare and the Allocation of Resources for Invention', in National Bureau of Economic Research, *The Rate and Direction of Inventive Activity*, Princeton, NJ: Princeton University Press.

Arrow, K. (1962b) 'The Economic Implications of Learning by Doing', *The Review of Economic Studies* 29(3): 155–173.

Arrow, K. and F. Hahn (1971) *General Competitive Analysis*, San Francisco: Holden-Day.

Arundel, A. (2000) 'Measuring the Economic Impacts of Biotechnology: From R&D to Applications', in J. de la Mothe and J. Niosi (eds) *Economics and Social Dynamics of Biotechnology*, London: Kluwer Academic Publishers.

Avorn, J. (2004) *Powerful Medicines: The Benefits, Risks, and Costs of Prescription Drugs*, New York: Knopf.

Bair, J. (2005) 'Global Capitalism and Commodity Chains: Looking Back, Going Forward', *Competition and Change* 9(2): 153–180.

Barnes, S.B. (1974) *Scientific Knowledge and Social Theory*, London: Routledge & Kegan Paul.

Barry, A. and D. Slater (2002) 'Introduction: The Technological Economy', *Economy & Society* 31(2): 175–193.

Barton, J. (2000) 'Intellectual Property Rights: Reforming the Patent System', *Science* 287: 1933–1934.

Bauer, M. and G. Gaskell (eds) (2002) *Biotechnology: The Making of a Global Controversy*, Cambridge: Cambridge University Press.

Beck, U. (1992) *Risk Society: Towards a New Modernity*, London: Sage.

Beck, U. (1999) *World Risk Society*, Cambridge: Polity.

Beck, U. (2006) *The Cosmopolitan Vision*, Cambridge: Polity.

Beck, U., A. Giddens and S. Lash (1994) *Reflexive Modernization: Politics, Tradition and Aesthetics in the Modern Social Order*, Palo Alto, CA: Stanford University Press.

Beddington, J. (2009) 'Food, Energy, Water and the Climate: A Perfect Storm of Global Events?' Speech at *Sustainable Development 09* Conference, London, 19 March 2009, available at www.dius.gov.uk/news_and_speeches/speeches/john_beddington/perfect-storm, accessed 11 October 2010.

Bekelman, J.E. *et al.* (2003) 'Scope and Impact of Financial Conflicts of Interest in Biomedical Research', *Journal of the American Medical Association* 289: 454–465.

Bell, D. (1974) *The Coming of Post-Industrial Society*, London: Heinemann.

Bellamy Foster, J. (2000) *Marx's Ecology: Materialism and Nature*, New York: Monthly Review Press.

Benton, T. (1981) 'Realism and Social Science: Some Comments of Roy Bhaskar's "The Possibility of Naturalism"', *Radical Philosophy* 27.

Benton, T. (ed.) (1996) *The Greening of Marxism*, New York: Guilford Press.

Benton, T. (2001) 'Environmental Sociology: Controversy and Continuity', *Sociologisk Tidsskrift (Journal of Sociology)* 9(1–2): 5–48.

Benton, T. (2003) 'Sociology and the Environment', in E. Page and J. Proops (eds), *Environmental Thought*, Cheltenham and Northampton, MA: Edward Elgar.

Berkhout, P. and C. van Bruchem (2008) *Agricultural Economic Report of The Netherlands*, LEI, Rapport 2008-030, The Hague: LEI.

Bernstein, H. (2001) '"The Peasantry" in Global Capitalism: Who, Where and Why?', *Socialist Register* 37: 25–51.

Bessen, J. and M. Meurer (2008) *Patent Failure: How Judges, Bureaucrats and Lawyers Put Innovators at Risk*, Princeton, NJ: Princeton University Press.

Bhaskar, R. (1986) *Scientific Realism and Human Emancipation*, London: Verso.

Bhaskar, R. (1993) *Dialectic: The Pulse of Freedom*, London: Verso.

Bhaskar, R. (1998) *The Possibility of Naturalism* (3rd edition), London and New York: Routledge.

Bhaskar, R. (2000) *From East to West*, London: Routledge.

Bhaskar, R. (2002) *Meta-Reality*, London, Thousand Oaks, CA and New Delhi: Sage.

Bhaskar, R. (2008) *A Realist Theory of Science* (2nd edition), London: Routledge.

Bieler, A. and A. Morton (2004) 'A Critical Theory Route to Hegemony, World Order and Historical Change: Neo-Gramscian Perspectives in International Relations', *Capital & Class* 82: 85–113.

Bijker, W., R. Bals and R. Hendricks (2009) *The Paradox of Scientific Authority: The Role of Scientific Advice in Democracies*, Cambridge, MA: MIT Press.

Birch, K. (2006) 'The Neoliberal Underpinnings of the Bioeconomy: The Ideological Discourses and Practices of Economic Competitiveness', *Genomics, Society and Policy* 2(3): 1–15.

Birch, K. and V. Mykhnenko (eds) (2010) *The Rise and Fall of Neoliberalism*, London: Zed Books.

Birch, K. and D. Tyfield (forthcoming) 'Theorizing the Bioeconomy: *Biovalue, Biocapital, Bioeconomics* or . . . What?', *Science, Technology and Human Values*.

Birch, K., L. Levidow and T. Papaioannou (2010) 'Sustainable Capital? The Neoliberalization of Nature and Knowledge in the European "Knowledge-Based Bio-Economy"', *Sustainability* 2: 2898–2918.

Birch, N. (2010) 'The Future Central Role of IPM in EU Crop Protection: How Can Ecological Research be Put into Practice?', Presentation to the Society of Experimental Botany Food Security and Safety International Symposium, Lancaster University, September 2010, available at www.sebiology.org/education/foodsecurity.html.

Blackburn, S. (2006) 'Finance's Fourth Dimension', *New Left Review* 39: 39–72.

Blumenstyk, G. (2007) 'Berkeley Professors Seek Voice in Research-Institute Deal with Energy Company', *Chronicle of Higher Education*, April 13, A33.

Blumenthal, D., M. Gluck, K. Seashore Louis, M. Stoto and D. Wise (1986a) 'Industry Support of University Research in Biotechnology', *Science* 231: 242–246.

Blumenthal, D., M. Gluck, K. Seashore Louis and D. Wise (1986b) 'University-Industry Research Relationships in Biotechnology: Implications for the University', *Science* 232: 1361–1366.

Bodenheimer, T. (2000) 'Uneasy Alliance: Clinical Investigators and the Pharmaceutical Industry', *New England Journal of Medicine* 342: 1539–1544.

Boldrin, M. and D. Levine (2010) *Against Intellectual Monopoly*, Cambridge: Cambridge University Press.

Boltanski, L. and E. Chiapello (2005) *The New Spirit of Capitalism*, London: Verso.

Boltanski, L. and L. Thévenot (2006) *On Justification: Economies of Worth* (trans. Catherine Porter), Princeton, NJ: Princeton University Press.

Bonanno, A., L. Busch, W. Friedland and L. Gouveia (eds) (1994) *From Columbus to ConAgra: Globalization of Agriculture and Food*, Lawrence, KS: University Press of Kansas.

Boyle, J. (1996) *Shamans, Software and Spleens: Law and the Construction of the Information Society*, Cambridge, MA and London: Harvard University Press.

Boyle, J. (2003) 'The Second Enclosure Movement and the Construction of the Public Domain', *Law and Contemporary Problems* 66: 33–73.

Brennan, T. (2000) *Exhausting Modernity – Grounds for a New Economy*, London and New York: Routledge.

Brenner, R. (2004) 'New Boom or New Bubble?', *New Left Review* 25: 57–100.

British Medical Journal (2003) 31 May edition, 1199.

Brock, W. and S. Durlauf (1999) 'A Formal Model of Theory Choice in Science', *Economic Theory* 14: 113–130, reprinted in P. Mirowski and E.-M. Sent (eds) (2002b) *Science Bought and Sold*, Chicago: University of Chicago Press.

Brown A., S. Fleetwood and J.M. Roberts (eds) (2002) *Critical Realism and Marxism*, London and New York: Routledge.

Brown, J.R. (2000) 'Privatizing the University – the New Tragedy of the Commons', *Science* 290: 1701.

Brown, M.B. (2010) 'Coercion, Corruption, and Politics in the Commodification of Academic Science', in H. Radder (ed.), *The Commodification of Academic Research: Science and the Modern University*, Pittsburgh: University of Pittsburgh Press.

Brown, N. and M. Michael (2003) 'A Sociology of Expectations: Retrospecting Prospects and Prospecting Retrospects', *Technology Analysis & Strategic Management* 15(1): 3–19.

Brown, N., B. Rappert and A. Webster (eds) (2000) *Contested Futures: A Sociology of Prospective Technoscience*, Aldershot: Ashgate.

Brueckner, A.L. (1996) 'Modest Transcendental Arguments', *Noûs* 30, Supplement: Philosophical Perspectives, 10, Metaphysics: 265–280.

Bud, R. (1992) 'The Zymotechnic Roots of Biotechnology', *British Journal for the History of Science* 25: 127–144.

Bud, R. (1993) *The Uses of Life: A History of Biotechnology*, Cambridge: Cambridge University Press.

Bud, R. (1998) 'Molecular Biology and the Long-Term History of Biotechnology', in A. Thackray (ed.), *Private Science: Biotechnology and the Rise of the Molecular Sciences*, Philadelphia: University of Pennsylvania Press.

Burawoy, M. (2005) 'For Public Sociology', *American Sociological Review* 70(1): 4–28.

Burkett, P. (1999) *Marx & Nature: A Red and Green Perspective*, Basingstoke: Palgrave Macmillan.

Busch, L. (2007) 'Performing the Economy, Performing Science: From Neoclassical to Supply Chain Models in the Agrifood Sector', *Economy & Society* 36(3): 437–466.

Busch, L. (2010) 'Acting Sustainably: Governance Through Standards in a Time of "Corporate Science"', Presentation to the Society of Experimental Botany Annual Conference, Lancaster, September 2010.

Busch, L. and C. Bain (2004) 'New! Improved? The Transformation of the Global Agrifood System', *Rural Sociology* 69(3): 321–346.

Bush, R. (2010) 'Food Riots: Poverty, Power and Protest', *Journal of Agrarian Change* 10(1): 119–129.

Bush, V. (1945) *Science: The Endless Frontier*, Washington, DC: US Government Printing Office.

Caffentzis, G. (2007) 'Crystals and Analytic Engines: Historical and Conceptual Preliminaries to a New Theory of Machines', *Ephemera* 7(1): 24–45.

Callon, M. (1994) 'Is Science a Public Good? Fifth Mullins Lecture, Virginia Polytechnic Institute, 23 March 1993', *Science, Technology and Human Values* 19(4): 395–424.

Callon, M. (1998a) 'Introduction: The Embeddedness of Economic Markets in Economics', in M. Callon (ed.), *The Laws of the Markets*, Oxford and Malden, MA: Blackwell.

Callon, M. (1998b) 'An Essay on Framing and Overflowing: Economic Externalities Revisited by Sociology', in M. Callon (ed.), *The Laws of the Markets*, Oxford and Malden, MA: Blackwell.

Callon, M. (2002) 'From Science as an Economic Activity to Socioeconomics of Scientific Research: The Dynamics of Emergent and Consolidated Techno-economic Networks', in P. Mirowski and E.-M. Sent (eds), *Science Bought and Sold*, Chicago: University of Chicago Press.

Callon, M., P. Lascoumes and Y. Barthe (2009) *Acting in an Uncertain World: An Essay on Technical Democracy* (trans. Graham Burchell), Cambridge, MA: MIT Press.

Calvert, J. (2004) 'The Idea of 'Basic Research' in Language and Practice', *Minerva* 45: 251–268.

Campbell, E., B. Claridge, M. Gokhale, L. Birenbaum, S. Hilgartner, N. Holtzman and D. Blumenthal (2002) 'Data Withholding in Academic Genetics: Evidence from a National Survey', *Journal of the American Medical Association* 287: 473–480.

Cao, C. (2004) 'Zhongguancun and China's High-Tech Parks in Transition: "Growing Pains" or "Premature Senility"?', *Asian Survey* 44(5): 647–668.

Carnoy, M. (1993a) 'Introduction', in M. Carnoy (ed.), *The New Global Economy in the Information Age: Reflections on our Changing World*, University Park, PA: Pennsylvania State University Press.

Carnoy, M. (1993b) 'Multinationals in a Changing World Economy: Whither the Nation-State?', in M. Carnoy (ed.), *The New Global Economy in the Information Age: Reflections on our Changing World*, University Park, PA: Pennsylvania State University Press.

Carter, B. and C. New (2004) *Making Realism Work: Realist Social Theory and Empirical Research*, London: Routledge.

Cassam, Q. (1987) 'Transcendental Arguments, Transcendental Synthesis and Transcendental Idealism', *The Philosophical Quarterly* 37(149): 355–378.

Castells, M. (1993) 'The Informational Economy and the New International Division of Labor', in M. Carnoy (ed.), *The New Global Economy in the Information Age: Reflections on our Changing World*, University Park, PA: Pennsylvania State University Press.

Castells, M. (1996) *The Rise of the Network Society*, Malden, MA and Oxford: Blackwell.

Castells, M. (1997) *The Power of Identity*, Malden, MA and Oxford: Blackwell.

Castells, M. (1998) *End of Millennium*, Malden, MA and Oxford: Blackwell.

Castree, N. (2002) 'False Antithesis? Marxism, Nature and Actor-Networks', *Antipode* 34(1): 111–146.

Castree, N. (2006) 'From Neoliberalism to Neoliberalisation: Consolations, Confusions and Necessary Illusions', *Environment & Planning A* 38(1): 1–6.

Castree, N. (2008) 'Neoliberalising Nature: The Logics of Deregulation and Reregulation', *Environment & Planning A* 40: 131–152.

Chang, H.-J. (2002) *Kicking Away the Ladder*, London: Anthem.

Chatham House (2009) *Food Futures: Rethinking UK Strategy*, London: Royal Institute of International Affairs (Chatham House).

Chatham House (2010) *Investing in Science – Securing Future Prosperity* Conference, Royal Institute for International Affairs, London.

Chen, C.-H. and H.-T. Shih (2005) *High-Tech Industries in China*, Cheltenham: Edward Elgar.

Chen, K. and M. Kenney (2007) 'Universities/ Research Institutes and Regional Innovation Systems', *World Development* 35(6): 1056–1074.

China Daily (2010) 'GM Grain Still "Long Distance Away"', *China Daily* 11 March 2010, available at www.chinadaily.com.cn/china/2010npc/2010-03/11/content_9570242.htm (accessed 5 May 2010).

Chouliaraki, L and N. Fairclough (1999) *Discourse in Late Modernity: Rethinking Critical Discourse Analysis*, Edinburgh: Edinburgh University Press.

Cochrane, W. (1993) *The Development of American Agriculture*, Minneapolis, MN: University of Minnesota Press.

Cohen, W. and D. Levinthal (1990) 'Absorptive Capacity: A New Perspective on Learning and Innovation', *Administrative Science Quarterly* 35: 128–152.

Cohen, W. and S. Merrill (eds) (2003) *Patents in the Knowledge-Based Economy*, Washington, DC: National Academy Press.

Cohen, W., R. Nelson and J.P. Walsh (2002) 'Links and Impacts: The Influence of Public Research on Industrial R&D', *Management Science* 48(1): 1–23.

Collini, S. (2010) 'Browne's Gamble: The Future of the Universities', *London Review of Books*, 4 November, 32(21): 23–25.

Collins, H.M. (1974) 'The TEA Set: Tacit Knowledge and Scientific Networks', *Social Studies of Science* 4: 165–186.

Collins, H.M. (1985/1992) *Changing Order: Replication and Induction in Scientific Practice*, London: Sage.

Collins, H.M. and R. Evans (2002) 'The Third Wave of Science Studies: Studies of Expertise and Experience', *Social Studies of Sciences* 32(2): 235–296.

Collinson, J.A. (2004) 2004 'Occupational Identity on the Edge: Social Science Contract Researchers in Higher Education', *Sociology* 38(2): 313–329.

Colyvas, J., M. Crow, A. Gelijns, R. Mazzoleni, R. Nelson, N. Rosenberg and B. Sampat (2002) 'How Do University Inventions Get Into Practice?', *Management Science* 48(1): 61–72.

Cooke, P., M. Heidenreich and H.-J. Braczyk (2004) *Regional Innovation Systems: The Role of Governances in a Globalized World*, London: Routledge.

Cooper, M. (2008) *Life as Surplus – Biotechnology and Capitalism in the Neoliberal Era*, Seattle, WA and London: University of Washington Press.

Couldry, N. and A. McRobbie (2010) 'Death of the University, English Style', *Culture Machine*, available at: www.culturemachine.net.

Cox, R. (1987) *Production, Power, and World Order*, New York and Guildford: Columbia University Press.

Cox, R. with T. Sinclair (1996) *Approaches to World Order*, Cambridge: Cambridge University Press.

Cruikshank, J. (2002) 'Critical Realism and Critical Philosophy: On the Usefulness of Philosophical Problems', *Journal of Critical Realism* 1: 50–66.

Dasgupta, P. (2002) 'Modern Economics and Its Critics', in U. Mäki (ed.), *Fact and Fiction in Economics: Models, Realism and Social Construction*, Cambridge: Cambridge University Press.

Dasgupta, P. and P. David (1987) 'Information Disclosure and the Economics of Science and Technology', in G. Feiwel (ed.), *Arrow and the Ascent of Modern Economic Theory*, New York: New York University Press.

Daston, L. (1992) 'Objectivity and the Escape from Perspective', *Social Studies of Science* 22(4): 597–618.

Daston, L. and P. Galison (2007) *Objectivity*, New York: Zone Books.

Datta, R., J. Frauley and F. Pearce (2010) 'Debate: Situation Critical – For a Critical, Reflexive, Realist, Emancipatory Social Science', *Journal of Critical Realism* 9(2): 227–247.

David, P. (1993) 'Intellectual Property Institutions and the Panda's Thumb: Patents, Copyrights and Trade Secrets in Economic Theory and History', in M. Wallerstein, M. Mogee and R. Schoen (eds), *Global Dimensions of Intellectual Property Rights in Science and Technology*, Washington, DC: National Academy Press.

David, P. (1998) 'Common Agency Contracting and the Emergence of Open Science Institutions', *American Economic Review: Papers and Proceedings* 88(2): 15–21.

David, P. and P. Dasgupta (1994) 'Toward a New Economics of Science', *Research Policy* 23: 487–521.

Davidsen, B. (2005) 'Critical Realism in Economics – a Different View', *Post-Autistic Economics Review* 33: 36–50.

Davies, W. (2010) 'The Rise of the Guilty Economist: Hybrid Policy Metrics and Governance', Presentation to the *After Markets* Workshop, Oxford, 23 April 2010.

Davis, G.F., M. Yoo, and W.E. Baker (2003) 'The Network Topography of the American Corporate Elite, 1982-2001', *Strategic Organisations* 1: 301–326.

De Angelis, M. (2004) 'Separating the Doing and the Deed: Capital and the Continuous Character of Enclosures', *Historical Materialism* 12(2): 57–87.

Deere, C. (2009) *The Implementation Game: The TRIPs Agreement and the Global Politics of Intellectual Property Reform in Developing Countries*, Oxford: Oxford University Press.

Demeritt, D. (2000) 'The New Social Contract for Science: Accountability, Relevance, and Value in US and UK Science and Research Policy', *Antipode* 32(3): 308–329.

Dennis, K. and J. Urry (2009) *After the Car*, Cambridge: Polity.

Dennis, M. (1987) 'Accounting for Research: New Histories of Corporate Laboratories and the Social History of American Science', *Social Studies of Science* 17(3): 479–518.

Dennis, M. (2004) 'Reconstructing Sociotechnical Order: Vannevar Bush and US Science Policy', in S. Jasanoff (ed.), *States of Knowledge*, London: Routledge.

Department of Food and Rural Affairs (DEFRA) (2009) *Food 2030*, London: DEFRA, available at www.defra.gov.uk/foodfarm/food/pdf/food2030strategy.pdf (accessed 20 February 2010).

Diamond, A. (1988) 'Science as a Rational Enterprise', *Theory and Decision* 24: 147–167.

Diamond, C. (1991) *The Realistic Spirit: Wittgenstein, Philosophy and the Mind*, Bradford Book, Cambridge, MA: MIT Press.

Dicken, P (2007) *Global Shift: Mapping the Contours of the World Economy* (5th edition), London: Sage.

Dicken, P., P. Kelly, K. Olds, and H. Yeung (2001) 'Chains and Networks, Territories and Scales: Towards a Relational Framework for Analysing the Global Economy', *Global Networks* 1(2): 89–112.

DiMasi, J., R. Hansen and H. Grabowski (2003) 'The Price of Innovation: New Estimates of Drug Development Costs', *Journal of Health Economics* 22: 151–185.

DiMasi, J., R. Hansen and H. Grabowski (2005a) 'Extraordinary Claims Require Extraordinary Evidence', *Journal of Health Economics* 24: 1034–1044.

DiMasi, J., R. Hansen and H. Grabowski (2005b) 'Setting the Record Straight on Setting the Record Straight: Response to the Light and Warburton Rejoinder', *Journal of Health Economics* 24: 1049–1053.

Directorate-General for Agriculture and Rural Development (2006) *Agricultural Trade Policy Analysis: Agricultural Commodity Markets – Past Developments and Outlook*, Brussels: EC.

Directorate-General for Enterprise (2009) *Taking Bio-Based from Promise to Market: Measures to Promote the Market Introduction of Innovative Bio-based Products*, European Commission, Brussels: DG Enterprise.

Directorate-General for Research (2005) *New Perspectives on the Knowledge-Based Bio-Economy: Conference Report*, European Commission, Brussels: DG Research.

Downward, P. (ed.) (2004) *Applied Economics and the Critical Realist Critique*, London: Routledge.

Drahos, P. (2007) '"Trust Me": Patent Offices in Developing Countries', *Centre for Governance of Knowledge and Development Working Paper*, Australia National University, Canberra.

Drahos, P. (2010) *The Global Governance of Knowledge: Patent Offices and their Clients*, Cambridge: Cambridge University Press.

Drahos, P. and J. Braithwaite (2002) *Information Feudalism: Who Owns the Knowledge Economy?*, London: Earthscan.

Drori, G., J. Meyer, F. Ramirez and E. Schofer (eds) (2003) *Science in the Modern World Polity: Institutionalization and Globalization*, Stanford, CA: Stanford University Press.

Duckett, J. (2001) 'Bureaucrats in Business, Chinese-Style', *World Development* 29(1): 23–37.

Dupré, J. (1995) *The Disorder of Things: Metaphysical Foundations of the Disunity of Science*, Cambridge, MA: Harvard University Press.

Dutfield, G. (2003) *Intellectual Property Rights and the Life Science Industries: A 20th Century History*, Aldershot: Ashgate.

The Economist (2005) 'Bayhing for Blood or Doling out the Cash?', Vol. 377, Issue 8458, 24 December.

The Economist (2009a) 'A Giant Falls', Vol. 391, Issue 8634, 4 June.

The Economist (2009b) 'The Other-Worldly Philosophers', Vol. 392, Issue 8640, 16 July.

The Economist (2010a) 'Investing in Brains', Vol. 394, Issue 8666, 21 January.

The Economist (2010b) 'Special Report on Innovation in Emerging Markets', Vol. 395, Issue 8678, 15 April.

The Economist (2010c) 'Special Report on the Human Genome', Vol. 395, Issue 8687, 17 June.

The Economist (2010d) 'Schools of Hard Knocks', Vol. 396, Issue 8699, 9 September.

The Economist (2010e) 'Learning the Right Lesson', Vol. 396, Issue 8699, 9 September.

The Economist (2010f) 'Grey-Sky Thinking', Vol. 396, Issue 8700, 16 September.

Eisenberg, R. (1987) 'Proprietary Rights and the Norms of Science in Biotechnology Research', *Yale Law Journal* 97: 177–231.

Eisenberg, R. (1996) 'Public Research and Private Development: Patents and Technology Transfer in Government-Sponsored Research', *Virginia Law Review* 82: 1663–1727.

Elman, B. (2007) 'New Directions in the History of Modern Science in China', *Isis* 98(3): 517–523.

Elson, D. (1979) 'The Value Theory of Labour', in *Value: The Representation of Labour in Capitalism*, London: CSE Books.

Epstein, S. (1996) *Impure Science: AIDS, Activism and the Politics of Knowledge*, Berkeley, CA: University of California Press.

Ernst, D. (2008) 'Innovation Offshoring and Asia's Electronics Industry: The New Dynamics of Global Networks', *International Journal of Technological Learning, Innovation & Development* 1(4): 551–576.

Etzkowitz, H. and L. Peters (1991) 'Profiting from Knowledge: Organisational Innovations and the Evolution of Academic Norms', *Minerva* 29(R): 133–166.

Etzkowitz, H., A. Webster, C. Gebhardt and B. Terra (2000) 'The Future of the University and the University of the Future: Evolution of Ivory Tower to Entrepreneurial Paradigm', *Research Policy* 29: 313–330.

Eun, J.-H., K. Lee and G. Wu (2006) 'Explaining the "University-Run Enterprises" in China', *Research Policy* 35(9): 1329–1346.

Euractiv (2004a) 'The Lisbon Agenda', available at www.euractiv.com/en/future-eu/lisbon-agenda/article-117510.

Euractiv (2004b) 'Growth and Jobs: Relaunch of the Lisbon Strategy', available at www.euractiv.com/en/innovation/growth-jobs-relaunch-lisbon-strategy/article-131891.

European Commission (2000) *The Lisbon European Council: An Agenda of Economic and Social Renewal in Europe*, DOC/00/7, Brussels: EC, available at http://ec.europa.eu/growthand jobs/pdf/lisbon_en.pdf.

European Commission (2002) *Life Sciences and Biotechnology: A Strategy for Europe*, Brussels: EC.

European Commission (2007) *Strategic Report on the Renewed Lisbon Strategy for Growth and Jobs: Launching the New Cycle (2008–2010): Keeping Up the Pace of Change*, COM(2007) 803, Brussels: EC, available at http://ec.europa.eu/growthandjobs/pdf/european-dimension-200712-annual-progress-report/200712-annual-report_en.pdf.

European Commission (2010) *Europe 2020: A Strategy for Smart, Sustainable and Inclusive Growth*, Brussels: EC.

Evidence (2007) *Briefing Document on Research Metrics in China*, Leeds: Evidence Ltd.

Fan, F. (2007) 'Redrawing the Map: Science in Twentieth-Century China', *Isis* 98(3): 524–538.

Fine, B. (1986) 'Introduction', in B. Fine (ed.), *The Value Dimension: Marx vs. Ricardo and Sraffa*, London and New York: Routledge & Kegan Paul.

Fine, B. (1999) 'A Question of Economics: Is It Colonizing the Social Sciences?', *Economy & Society* 28(3): 403–425.

Fine, B. (2010) 'Zombieconomics: The Living Death of the Dismal Science', in K. Birch and V. Mykhnenko (eds), *The Rise and Fall of Neoliberalism*, London: Zed Books.

Fine, B. and D. Milonakis (2009) *From Economics Imperialism to Freakonomics: The Shifting Boundaries Between Economics and Other Social Sciences*, London and New York: Routledge.

Fine, B. and A. Saad-Filho (2004) *Marx's 'Capital'* (4th edition), London and Sterling, VA: Pluto Press.

Finlayson, A. (2010) 'Britain, Greet the Age of Privatised Higher Education', *OpenDemocracy*, 9 December, available at: www.opendemocracy.net/ourkingdom/alan-finlayson/britain-greet-age-of-privatised-higher-education.

Fischer, W. and M. von Zedtwitz (2004) 'Chinese R&D: Naissance, Renaissance or Mirage?', *R&D Management* 34(4): 349–365.

Fleetwood, S. (ed.) (1999) *Critical Realism in Economics: Development and Debate*, London: Routledge.

Fleetwood, S. (2001) 'What Kind of *Theory* is Marx's Labour *Theory* of Value? A Critical Realist Inquiry', *Capital & Class* 73: 41–77, reprinted in A. Brown, S. Fleetwood and J.M. Roberts (eds) (2002), *Critical Realism and Marxism*, London: Routledge.

Food and Agriculture Organisation (FAO) (2010) *The State of Food and Agriculture 2009*, Rome: FAO.

Foray, D. (2004) *The Economics of Knowledge*, Cambridge, MA: MIT Press.

Framework Programme 7 (2010) 'Food, Agriculture and Fisheries, and Biotechnology', available at http://cordis.europa.eu/fp7/kbbe/home_en.html.

Frauley, J. (2007) 'Towards an Archaeological-Realist Foucauldian Analytics of Government', *British Journal of Criminology* 47: 617–633.

Frey, K. (1996) *National Plant Breeding Study – 1: Human and Financial Resources Devoted to Plant Breeding Research and Development in the United States in 1994*, Ames, IA: Iowa State University.

Friedmann, H. (2005) 'Feeding the Empire: The Pathologies of Globalized Agriculture', *Socialist Register* 41: 124–143.

Fuglie, K., N. Ballenger, K. Day, C. Klotz, M. Ollinger, J. Reilly, U. Vasavada and J. Yee (1996) 'Agricultural Research and Development: Public and Private Investments Under Alternative Markets and Institutions', *US Department of Agriculture, Agricultural Research and Development, Agricultural Economics Report No.* 735, Washington, DC: DoA.

Fullbrook, E. (2003) *The Crisis in Economics*, London: Routledge.

Fullbrook, E. (ed.) (2008) *Ontology and Economics: Tony Lawson and His Critics*, London: Routledge.

Fuller, S. (1994) 'Mortgaging the Farm to Save the (Sacred) Cow', *Studies in the History and Philosophy of Science* 25: 251–261.

Fuller, S. (2010) 'Capitalism and Knowledge: The University Between Commodification and Entrepreneurship', in H. Radder (ed.), *The Commodification of Academic Research: Science and the Modern University*, Pittsburgh, PA: University of Pittsburgh Press.

Gabriele, A. (2002) 'S&T Policies and Technical Progress in China's Industry', *Review of International Political Economy* 9(2): 333–373.

Galison, P. (1987) *How Experiments End*, Chicago: University of Chicago Press.

Galison, P. and D. Stump (eds) (1996) *The Disunity of Science: Boundaries, Contexts and Power*, Stanford, CA: Stanford University Press.

Gallini, N.T. (2002) 'The Economics of Patents: Lessons from Recent U.S. Patent Reform', *The Journal of Economic Perspectives* 16(2): 131–154.

Geiger, R. (2004) *Knowledge and Money*, Stanford, CA: Stanford University Press.

Gereffi, G., J. Humphrey and T. Sturgeon (2005) 'The Governance of Global Value Chains', *Review of International Political Economy* 12(1): 78–104.

Gibbons, M., C. Limoges, H. Nowotny, S. Schwartzman, P. Scott and M. Trow (1994) *The New Production of Knowledge: The Dynamics of Science and Research in Contemporary Societies*, London, Thousand Oaks, CA and New Delhi: Sage.

Gill, S. (1993a) 'Gramsci and Global Politics: towards a Post-Hegemonic Research Agenda', in S. Gill (ed.), *Gramsci, Historical Materialism and International Relations*, Cambridge: Cambridge University Press.

Gill, S. (1993b) 'Epistemology, Ontology and the "Italian school"', in S. Gill (ed.), *Gramsci, Historical Materialism and International Relations*, Cambridge: Cambridge University Press.

Glennie, P. and N. Thrift (1996) 'Reworking E. P. Thompson's "Time, Work-Discipline and Industrial Capitalism"', *Time and Society* 5(3): 275–299.

Glyn, A. (2006) *Capitalism Unleashed*, Oxford: Oxford University Press.

Godin, B. (2006) 'The Knowledge-Based Economy: Conceptual Framework or Buzzword?', *Journal of Technology Transfer* 31: 17–30.

Goldman, A. (1986) *Epistemology and Cognition*, Cambridge, MA: Harvard University Press.

Goldman, A. and M. Shaked (1991) 'An Economic Model of Scientific Activity and Truth Acquisition', *Philosophical Studies* 63: 31–55.

Goldthorpe, J.H. (2010) 'Analysing Social Inequality: A Critique of Two Recent Contributions from Economics and Epidemiology', *European Sociological Review* 26(6): 731–744.

Goodman, D. (2001) 'Ontology Matters: The Relational Materiality of Nature and Agri-Food Studies', *Sociologia Ruralis* 41(2): 182–200.

Goozner, M. (2004) *The $800 Million Pill: The Truth Behind the Cost of New Drugs*, Berkeley, CA: University of California Press.

Gramsci, A. (1971) *Selections from the Prison Notebooks*, London: Lawrence & Wishart.

Gregory, P. (2010) 'Feeding 9 Billion: The Challenge to Sustainable Crop Production', Presentation to the Society of Experimental Botany Food Security and Safety International Symposium, Lancaster University, September 2010, available at www.sebiology.org/education/foodsecurity.html.

Gross, M. (2010) *Ignorance and Surprise: Science, Society and Ecological Design*, Cambridge, MA: MIT Press.

Guan, J.C., R. Yam and C.K. Mok (2005) 'Collaboration between Industry and Research Institutes/Universities on Industrial Innovation in Beijing, China', *Technology Analysis & Strategic Management* 17(3): 339–353.

Hacker, J. and P. Pierson (2011) *Winner-Take-All Politics: How Washington Made the Rich Richer – and Turned Its Back on the Middle Class*, New York: Simon & Schuster.

Hacking, I. (1983) *Representing and Intervening*, Cambridge: Cambridge University Press.

Hahn, F. (1992) 'Answer to Backhouse: Yes', *Royal Economic Society Newsletter* 78: 5.

Hands, D.W. (1994) 'The Sociology of Scientific Knowledge, Some thoughts on the possibilities', in R. Backhouse (ed.), *New Directions in Economic Methodology*, London: Routledge.

Hands, D.W. (1997) 'Caveat Emptor: Economics and Contemporary Philosophy of Science', *Philosophy of Science* 64 (Proceedings): S107–S116.

Hands, D.W. (1999) 'Empirical Realism as Meta-Method – Tony Lawson on Neoclassical Economics', in S. Fleetwood (ed.), *Critical Realism in Economics: Development and Debate*, London: Routledge.

Hands, D.W. (2001) *Reflection without Rules: Economic Methodology and Contemporary Science Theory*, Cambridge: Cambridge University Press.

Harcourt, G. (2010) 'The Crisis in Mainstream Economics', *Real-World Economics Review* 53: 47–51.

Hardin, G. (1968) 'The Tragedy of the Commons', *Science*, 13 December, 162: 1243–1248.

Hartwig, M. (2009) ' "Orthodox" Critical Realism and the Critical Realist Embrace', *Journal of Critical Realism* 8(2): 233–257.

Harvey, D. (1982) *Limits to Capital*, Oxford: Basil Blackwell.

Harvey, D. (1996) *Justice, Nature and the Geography of Difference*, New York: Wiley-Blackwell.

Harvey, D. (2003) *The New Imperialism*, Oxford: Oxford University Press.

Hassan, G. (2010) 'The Fantasyland of "The Spirit Level" and the Limitations of the Health and Well-Being Industry', *OpenDemocracy*, 1 August, available at www.opendemocracy. net/ourkingdom/gerry-hassan/fantasyland-of-%E2%80%98-spirit-level%E2%80%99-and-limitations-of-health-and-well-being-indu.

Hausman, D. (1998) 'Problems with Realism in Economics', *Economics and Philosophy* 14: 185–213.

Hausman, D. (1999) 'Ontology and Methodology in Economics', *Economics and Philosophy* 15: 283–288.

Hayek, F. (1937) 'Economics and Knowledge', *Economica* 4(13): 33–54.

Hayek, F. (1945) 'The Use of Knowledge in Society', *American Economic Review* 35(4): 519–530.

Heller, M. and R. Eisenberg (1998) 'Can Patents Deter Innovation? The Anti-Commons in Biomedical Research', *Science* 280: 698–701.

Hempel, C. (1962) 'Explanation in Science and in History', in R. Colodny (ed.), *Frontiers of Science and Philosophy*, Pittsburg, PA: University of Pittsburgh Press.

Henderson, M. (2010) 'Problems with Peer Review', *The Lancet* 340: 1409.

Hicks, J. (1995) 'Published Papers, Tacit Competences and Corporate Management of the Public/Private Character of Knowledge', *Industrial and Corporate Change* 4(2): 401–424.

Hintikka, J. (1972) 'Transcendental Arguments: Genuine and Spurious', *Noûs* 6(3): 274–281.

Hodgson, G. (2004) 'On the Problem of Formalism in Economics', *Post-Autistic Economics Review* 28: 3–12.

Hong, W. (2007) 'Decline of the Centre', *Research Policy* 37: 580–595.

Hu, M.-C. and J. Matthews (2008) 'China's National Innovative Capacity', *Research Policy* 37: 1465–1479.

Huang, J., Q. Wang, Y. Zhang and J.F. Zepeda (2001) 'Agricultural Biotechnology Research Indicators: China', *Center for Chinese Agricultural Policy Working Paper*. Beijing: CAS.

Huang, J., S. Rozelle, C. Pray and Q. Wang (2002) 'Plant Biotechnology in China', *Science*, 25 January, 295: 674–677.

Huang, J., R. Hu, C. Pray and S. Rozelle (2004) 'Plant Biotechnology in China: Public Investments and Impacts on Farmers', *New Directions for a Diverse Planet*. Proceedings of the 4th International Crop Science Congress, 26 September–1 October 2004, Brisbane, Australia.

Hung, H. (2009) 'America's Head Servant? The PRC's Dilemma in the Global Crisis', *New Left Review* 60: 5–25.

Huws, U. (1999) 'Material World: The Myth of the Weightless Economy', *Socialist Register* 35: 30–55.

Hymer, S. (1975) 'The Multinational Corporation and the Law of Uneven Development', in H. Radice (ed.), *International Firms and Modern Imperialism – Selected Readings*, Harmondsworth: Penguin Books.

Hymer, S. (1979) *The Multinational Corporation: A Radical Approach: Papers by Stephen Herbert Hymer*, Robert B. Cohen, Nadine Felton, Morley Nkosi and Jaap van Liere (eds), Cambridge University Press: Cambridge.

International Panel on Climate Change (IPCC) (2007) *Climate Change 2007: Synthesis Report*, Geneva: IPCC.

Irwin, A., H. Rothstein, S. Yearley and E. McCarthy (1997) 'Regulatory Science – Towards a Sociological Framework', *Futures* 29(1): 17–31.

Jackson, T. (2009) *Prosperity without Growth?*, London: Sustainable Development Commission.

Jakobson, L. (2007) 'China Aims High in Science and Technology', in L. Jakobson (ed.), *Innovation with Chinese Characteristics*, Basingstoke: Palgrave Macmillan.

Jaffe, A. (2000) 'The U.S. Patent System in Transition: Policy Innovation and the Innovation Process', *Research Policy* 29: 531–557.

Jaffe, A. and J. Lerner (2007) *Innovation and Its Discontents*, Princeton, NJ: Princeton University Press.

Jasanoff, S. (ed.) (2004) *States of Knowledge: The Co-Production of Science and Social Order*, London: Routledge.

Jasanoff, S. (2005) *Designs on Nature*, Princeton, NJ: Princeton University Press.

Jasanoff, S. (2007) 'Technologies of Humility', *Nature*, 1 November, 450: 33.

Jasanoff, S. (2010) 'A New Politics of Innovation', in J. Pugh (ed.), *What Is Radical Politics Today*, Basingstoke: Palgrave Macmillan.

Jessop, B. (1997) 'Capitalism and Its Future: Remarks on Regulation, Government and Governance', *Review of International Political Economy* 4(3): 561–581.

Jessop, B. (1999) 'Reflections on Globalisation and Its (Il)logic(s)', in K. Olds, P. Kelly, L. Kong and H. Yeung (eds), *Globalization and the Asia Pacific: Contested Territories*, London and New York: Routledge.

Jessop, B. (2000a) 'The Crisis of the National Spatio-Temporal Fix and the Tendential Ecological Dominance of Globalizing Capitalism', *International Journal of Urban and Regional Research* 24(2): 323–360.

Jessop, B. (2000b) 'The State and the Contradictions of the Knowledge-Driven Economy', in J. Bryson, P. Daniels, N. Henry and J. Pollard (eds), *Knowledge, Space, Economy*, London and New York: Routledge.

Jessop, B. (2001) 'Beyond Developmental States: A Regulationist and State-Theoretical Analysis', Mimeo, Lancaster University.

Jessop, B. (2002a) *The Future of the Capitalist State*, Cambridge: Polity.

Jessop, B. (2002b) 'Capitalism, the Regulation Approach, and Critical Realism', in A. Brown, S. Fleetwood and J.M. Roberts (eds), *Critical Realism and Marxism*, London and New York: Routledge.

Jessop, B. (2005) 'Cultural Political Economy, the Knowledge-Based Economy, and the State', in A. Barry and D. Slater (eds), *The Technological Economy*, Abingdon: Routledge.

Jessop, B. (2007) 'Knowledge as Fictitious Commodity: Insights and Limits of a Polanyian Perspective', in A. Bugra and K. Agartan (eds), *Reading Karl Polanyi for the 21st Century: The Market Economy as a Political Project*, Basingstoke: Palgrave Macmillan.

Jessop, B., N. Fairclough and R. Wodak (2008) *Education and the Knowledge-Based Economy in Europe*, Rotterdam: Sense Publishers.

Jessop, B. and N.L. Sum (2006) *Beyond the Regulation Approach: Putting Capitalist Economies in their Place*, Cheltenham: Edward Elgar.

Jia, H., K.S. Jayaraman and S. Louët (2004) 'China Ramps Up Efforts to Commercialize GM Rice', *Nature Biotechnology* 22(6): 642.

Johnson, B., E. Lorenz and B.-Å. Lundvall (2002) 'Why all this Fuss about Codified and Tacit Knowledge?', *Industrial and Corporate Change* 11(2): 245–262.

Joseph, J. (2002) *Hegemony*, London: Routledge.

Joseph, J. and J.M. Roberts (2005) 'Derrida, Foucault and Zizek: Being Realistic About Social Theory', *New Formations* 56: 109–120.

Journal of Critical Realism (2009) *Special Issue on Causal Powers* 8(3).

Kaidesoja, T. (2005) 'The Trouble with Transcendental Arguments', *Journal of Critical Realism* 4(1): 28–61.

Kaidesoja, T. (2006) 'How Useful Are Transcendental Arguments for Critical Realist Ontology? A Response to Morgan', *Journal of Critial Realism* 5(2): 344–353.

Kant, I. (1781/1953) *The Critique of Pure Reason* (trans. Norman Kemp Smith), London: Macmillan.

Kassirer, J.P. (2004) *On the Take: How Medicine's Complicity with Big Business Can Endanger Your Health*, Oxford: Oxford University Press.

Kay, J. (2009) 'The Spirit Level: Review', *Financial Times*, 23 March.

Kay, L. (1993) *The Molecular Vision of Life: Caltech, the Rockefeller Foundation and the Rise of the New Biology*, New York, Oxford: Oxford University Press.

Kay, L. (1998) 'Problematizing Basic Research in Molecular Biology', in A. Thackray (ed.), *Private Science: Biotechnology and the Rise of the Molecular Sciences*, Philadelphia: University of Pennsylvania Press.

Kemp, S. (2005) 'Critical Realism and the Limits of Philosophy', *European Journal of Social Theory* 8(2): 171–191.

Kenney, M. (1986) *Biotechnology: The University-Industrial Complex*, London and New Haven, CT: Yale University Press.

Kenney, M. (1998) 'Biotechnology and the Creation of a New Economic Space', in A. Thackray (ed.), *Private Science: Biotechnology and the Rise of the Molecular Sciences*, Philadelphia: University of Pennsylvania Press.

Kincaid, H. (1997) 'Individualism and Rationality', in *Individualism and the Unity of Science*, Lanham, MD: Rowman & Littlefield.

Kincaid, H. and D. Ross (eds) (2009) *Oxford Handbook of Philosophy of Economics*, Oxford: Oxford University Press.

Kitcher, P. (1993) *The Advancement of Science: Science without Legend, Objectivity Without Illusions*, Oxford: Oxford University Press.

Kitcher, P. (2000) 'Reviving the Sociology of Science', *Philosophy of Science* 67: S33–S44.

Kleinman, D.L. (2003) *Impure Cultures – University Biology and the World of Commerce*, Madison, WI: University of Wisconsin Press.

Kleinman. D.L. (2010) 'The Commercialization of Academic Culture and the Future of the University', in H. Radder (ed.), *The Commodification of Academic Research: Science and the Modern University*, Pittsburgh, PA: University of Pittsburgh Press.

Kleinman, D. and S. Vallas (2001) 'Science, Capitalism, and the Rise of the "Knowledge Worker": The Changing Structure of Knowledge Production in the United States', *Theory and Society* 30: 451–492.

Klevorick, A., R. Levin, R. Nelson and S. Winter (1995) 'On the Sources and Significance of Interindustry Differences in Technological Opportunities', *Research Policy* 24(2): 185–205.

Kloppenburg, J. (2004) *First the Seed: The Political Economy of Plant Biotechnology*, 2nd edition, Madison, WI: University of Wisconsin Press.

Knorr-Cetina, K. (1981) *The Manufacture of Knowledge: An Essay on the Constructivist and Contextual Nature of Science*, Oxford and New York: Pergamon Press.

Kohler, R. (1991) *Partners in Science: Foundations and Natural Scientists, 1900–1945*, Chicago: University of Chicago Press.

Kortum, S. and J. Lerner (1999) 'What is Behind the Recent Surge in Patenting?', *Research Policy* 28: 1–22.

Krige, J. (2008) *American Hegemony and the Postwar Reconstruction of Science in Europe*, Cambridge, MA: MIT Press.

Krimsky, S. (2003) *Science in the Private Interest: Has the Lure of Profits Corrupted Biomedical Research?*, Lanham: Rowman & Littlefield.

Kroll, H. and I. Liefner (2008) 'Spin-off Enterprises as a Means of Technology Commercialisation in a Transforming Economy', *Technovation* 28: 298–313.

Kuemmerle, W. (1999) 'Foreign Direct Investment in Industrial Research in the Pharmaceutical and Electronics Industries – Results from a Survey of Multinational Firms', *Research Policy* 28: 179–193.

Lang, T., D. Barling and M. Caraher (2009) *Food Policy: Integrating Health, Environment and Society*, Oxford: Oxford University Press.

Latour, B. (1987) *Science in Action: How to Follow Scientists and Engineers Through Society*, Cambridge, MA: Harvard University Press.

Latour, B. (1993) *We Have Never Been Modern*, Hemel Hempstead: Harvester Wheatsheaf.

Latour, B. (1999) *Pandora's Hope: Essays on the Reality of Science Studies*, Cambridge, MA: Harvard University Press.

Latour, B. (2004) *Politics of Nature: How to Bring the Sciences into Democracy* (trans. Catherine Porter), Cambridge, MA: Harvard University Press.

Latour, B. and S. Woolgar (1986) *Laboratory Life*, Princeton, NJ: Princeton University Press.

Lave, R., P. Mirowski and S. Randalls (eds) (2010) Special Issue on 'STS and Neoliberal Science', *Social Studies of Science* 40(5).

Law, J. (ed.) (1991) *A Sociology of Monsters*, London: Routledge.

Lawrence, G., K. Lyon, T. Wallington (2010) 'Introduction: Food Security, Nutrition and Sustainability in a Globalized World', in G. Lawrence, K. Lyon, and T. Wallington (eds), *Food Security, Nutrition and Sustainability*, London: Earthscan.

Lawson, T. (1997) *Economics and Reality*, London and New York: Routledge.

Lawson, T. (1999a) 'What has Realism Got to Do With It?', *Economics and Philosophy* 15: 269–282.

Lawson, T. (1999b) 'Developments in *Economics as Realist Social Theory*', in S. Fleetwood (ed.), *Critical Realism in Economics: Development and Debate*, London: Routledge.

Lawson, T. (1999c) 'Critical Issues in *Economics as Realist Social Theory*', in S. Fleetwood (ed.), *Critical Realism in Economics: Development and Debate*, London: Routledge.

Lawson, T. (2003) *Reorienting Economics*, London and New York: Routledge.

Leach, M., I. Scoones and B. Wynne (eds) (2005) *Science and Citizens: Globalization and the Challenge of Engagement*, London and New York: Sage.

Leadbeater, C. and J. Wilsdon (2007) *The Atlas of Ideas*, London: Demos.

Lee, B., I. Iliev and F. Preston (2009) *Who Owns Our Low Carbon Future? Intellectual Property and Energy Technologies*, London: Royal Institute of International Affairs (Chatham House).

Lessig, L. (2001) *The Future of Ideas*, New York: Random House.

Lessig, L. (2002) 'The Architecture of Innovation', *Duke Law Journal* 51: 1783–1801.

Lewis, P. (ed.) (2004a) *Transforming Economics: Perspectives on the Critical Realist Project*, London: Routledge.

Lewis, P. (2004b) 'Transforming Economics? On Heterodox Economics and the Ontological Turn in Economic Methodology', in P. Lewis (ed.), *Transforming Economics: Perspectives on the Critical Realist Project*, London: Routledge.

Levidow, L., J. Murphy and S. Carr (2007) 'Recasting "Substantial Equivalence": Transatlantic Governance of GM Food', *Science, Technology & Human Values* 32: 26–64.

Levin, R., A. Klevorick, R. Nelson and S. Winter (1987) 'Appropriating the Returns from Industrial Research and Development', *Brookings Papers on Economic Activity* 3: 783–831.

Liebenau, J. (1984) 'Industrial R&D in Pharmaceutical Firms in the Early Twentieth Century', *Business History* 26: 329–346.

Lieberthal, K. (1992) 'Introduction: The "Fragmented Authoritarianism" Model and its Limitations', in K. Lieberthal and A. Thurston (eds), *Bureaucracy, Politics and Decision Making in Post-Mao China*, Berkeley, CA: University of California Press.

Light, D. and R. Warburton (2005a) 'Extraordinary Claims Require Extraordinary Evidence', *Journal of Health Economics* 24: 1030–1033.

Light, D. and R. Warburton (2005b) 'Setting the Record Straight in the Reply by DiMasi, Hansen and Grabowski', *Journal of Health Economics* 24: 1045–1048.

Liu, X. and S. White (2001) 'Comparing Innovation Systems: A Framework and Application to China's Transitional Context', *Research Policy* 30: 1091–1114.

Lukacs, G. (1972) *History and Class Consciousness*, Cambridge, MA: MIT Press.

Lundvall, B.-Å. (1996) 'The Social Dimension of The Learning Economy', DRUID Working Paper, available at www3.druid.dk/wp/19960001.pdf.

MacDonald, S. (2002) 'Exploring the Hidden Costs of Patents', in P. Drahos and R. Mayne (eds), *Global Intellectual Property Rights: Knowledge, Access and Development*, Basingstoke: Palgrave Macmillan.

Macnaghten, P. and J. Urry (1998) *Contested Natures*, London, Thousand Oaks, CA and New Delhi: Sage.

Magdoff, F., J.B. Foster and F. Buttel (2000) *Hungry for Profit*, New York: Monthly Review Books.

Majka, L. and T. Majka (2000) 'Organizing US Farm Workers: A Continuous Struggle', in F. Magdoff, J.B. Foster and F. Buttel (eds), *Hungry for Profit*, New York: Monthly Review Books.

Mäki, U. (2004) 'Economic Epistemology: Hopes and Horrors', *Episteme* 3: 211–222.

Mann, S. and J. Dickinson (1978) 'Obstacles to Development of a Capitalist Agriculture', *Journal of Peasant Studies* 5: 466–481.

Mansfield, E. (1986) 'Patents and Innovation: An Empirical Study', *Management Science* 32(2): 173–181.

Marsden, T. (2003) *The Condition of Rural Sustainability*, Assen: Royal Van Gorcum.

Marx, K. (1992) *Grundrisse*, Harmondsworth: Penguin.

Marx, K. (1999) *Capital*, abridged by D. McLellan, Oxford: Oxford University Press.

Mascarenhas, M. and L. Busch (2006) 'Seeds of Change: Intellectual Property Rights, Genetically Modified Soybeans and Seed Saving in the United States', *Sociologia Ruralis* 46(2): 122–138.

Maskus, K. (2000) *Intellectual Property Rights in the Global Economy*, Washington, DC: Institute for International Economics.

Matthews, D. (2002) *Globalising Intellectual Property Rights: The TRIPs Agreement*, London: Routledge.

May, C. (2000) *A Global Political Economy of Intellectual Property Rights*, London and New York: Routledge.

May, C. (2006) 'The Denial of History: Reification, Intellectual Property Rights and the Lessons of the Past', *Capital & Class* 88: 33–56.

May, C. and S. Sell (2006) *Intellectual Property Rights: A Critical History*, Boulder, CO and London: Lynne Rienner Publishers.

Mazzoleni, R. and R. Nelson (1998) 'Economic Theories about the Benefits and Costs of Patents', *Journal of Economic Issues* 32(4): 1031–1052.

McMichael, P. (2009) 'A Food Regime Analysis of the World Food Crisis', *Agriculture and Human Values* 4: 281–295.

Mellon, M., C. Benbrook and K. Lutz Benbrook (2001) *Hogging It! Estimates of Antimicrobial Abuse in Livestock*, Washington, DC: Union of Concerned Scientists.

Merges, R. (1999) 'As Many as Six Impossible Patents before Breakfast: Property Rights for Business Concepts and Patent System Reform', *Berkeley Technology Law Journal* 14: 577–615.

Merges, R. and R. Nelson (1990) 'On the Complex Economics of Patent Scope', *Columbia Law Review* 90(4): 839–916.

Merton, R.K. (1973) *The Sociology of Science: Theoretical and Empirical Investigations*, Chicago: Chicago University Press.

Miller, D. (2010) 'How Neoliberalism Got Where It Is: Elite Planning, Corporate Lobbying and the Release of the Free Market', in K. Birch and V. Mykhnenko (eds), *The Rise and Fall of Neoliberalism*, London: Zed Books.

Ministry of Agriculture of China (1990) *The Guideline for the Development of Science and Technology in Middle and Long Term: 1990–2000*, Beijing: MOA Press.

Ministry of Science and Technology of China (2007) *China Science and Technology Statistics Data Book*, Beijing: MOST.

Minogue, M. (1993) 'Theory and Practice in Public Policy and Administration', in Michael Hill (ed.), *The Policy Process: A Reader*, Harlow: Pearson, pp. 10–34.

Mirowski, P. (1995) 'Philip Kitcher's *Advancement of Science*: A Review Article', *Review of Political Economy* 7: 227–241.

Mirowski, P. (1996) 'The Economic Consequences of Philip Kitcher', *Social Epistemology* 10: 153–169.

Mirowski, P. (2009) 'Why There Is (as Yet) No Such Thing as an Economics of Knowledge', in H. Kincaid and D. Ross (eds), *Oxford Handbook of Philosophy of Economics*, Oxford: Oxford University Press.

Mirowski, P. (2011) *Science-Mart: Privatizing American Science*, Cambridge, MA: Harvard University Press.

Mirowski, P. and D. Plehwe (eds) (2009) *The Road to Mont Pelerin: The Making of the Neoliberal Thought Collective*, Cambridge, MA: Harvard University Press.

Mirowski, P. and E.-M. Sent (2002a) 'Introduction', in P. Mirowski and E.-M. Sent (2002b).

Mirowski. P. and E.-M. Sent (eds) (2002b) *Science Bought and Sold*, Chicago: University of Chicago Press.

Mirowski, P. and E.-M. Sent (2008) 'The Commercialization of Science and the Response of STS', in E. Hackett, J. Wacjman, O. Amsterdamska and M. Lynch (eds), *New Handbook of STS*, Cambridge (MA): MIT Press.

Mirowski, P. and R. Van Horn (2005) 'The Contract Research Organization and the Commercialization of Scientific Research', *Social Studies of Science* 35(4): 503–548.

Mohun, S. (2003) 'Does All Labour Create Value', in A. Saad-Filho (ed.), *Anti-Capitalism: A Marxist Introduction*, London and Sterling, VA: Pluto Press.

Motohashi, K. (2008) 'Assessment of Technological Capability in Science Industry Linkage in China by Patent Database', *World Patent Information* 30: 225–232.

Moulaert, F. and F. Sekia (2003) 'Territorial Innovation Models: A Critical Survey', *Regional Studies* 37(3): 289–302.

Mowery, D., R. Nelson, B. Sampat and A. Ziedonis (2001) 'The Growth of Patenting and Licensing by U.S. Universities: An Assessment of the Effects of the Bayh-Dole act of 1980', *Research Policy* 30: 99–119.

Mowery, D., R. Nelson, B. Sampat and A. Ziedonis (2004) *Ivory Tower and Industrial Innovation*, Stanford, CA: Stanford University Press.

Moynihan, R. and A. Cassels (2005) *Selling Sickness: How the World's Biggest Pharmaceutical Companies Are Turning Us All into Patients*, New York: Nation Books.

Murdoch, J. (1997) 'Inhuman/Nonhuman/Human', *Environment and Planning D: Society and Space* 15: 731–756.

Nagel, T. (1989) *The View from Nowhere*, Oxford and New York: Oxford University Press.

Narin, F., K.S. Hamilton and D. Olivastro (1997) 'The Increasing Link between US Technology and Public Science', *Research Policy* 26(3): 317–330.

National Institute for Health Care Management (NIHCM) (2002) 'Changing Patterns of Pharmaceutical Innovation', *The National Institute for Health Care Management Research and Educational Foundation*, Washington, DC.

National Science Foundation (2006) *Academic Research and Development Expenditures: Fiscal Year 2004*, NSF 06-323, Project Officer, Ronda Britt, Arlington, VA.

National Science Foundation (2010) *National Patterns of R&D Resources: 2008 Data Update*, NSF 10-314, Arlington, VA.

Naughton, B. (2007) *The Chinese Economy*, Cambridge, MA: MIT Press.

Nelson, R. (1959) 'The Simple Economics of Basic Scientific Research', *Journal of Political Economy* 67: 297–306.

Nelson, R. (2001) 'Observations on the Post-Bayh-Dole Rise of Patenting at American Universities', *Journal of Technology Transfer* 26: 13–19.

Nelson, R. (2004) 'The Market Economy, and the Scientific Commons', *Research Policy* 33: 455–471.

Newfield, C. (2003) *Ivy and Industry: Business and the Making of the American University, 1880–1980*, Durham, NC and London: Duke University Press.

Nierenberg, D. and L. Mastny (2005) *Happier Meals: Rethinking the Global Meat Industry*, Washington, DC: Worldwatch Institute.

Nightingale, P. (2004) 'The Myth of the Biotech Revolution', *Trends in Biotechnology* 22(11): 564–569.

Nixon, J., A. Marks, A. Rowland and M. Walker (2001) 'Towards a New Academic Professionalism: A Manifesto of Hope', *British Journal of Sociology of Education* 22(2): 227–244.

Noble, D. (2002) 'Digital Diploma Mills: The Automation of Higher Education', in P. Mirowski and E.M. Sent (eds), *Science Bought and Sold*, Chicago: University of Chicago Press.

Nolan, P. (2004) *China at the Crossroads*, Cambridge: Polity.

Nonaka, I. (1994) 'A Dynamic Theory of Organizational Knowledge Creation', *Organization Science* 5(1): 14–37.

Nonini, D. (2008) Is China Becoming Neoliberal? *Critique of Anthropology* 28(2): 145–176.

Nowotny, H., P. Scott and M. Gibbons (2002) *Re-Thinking Science*, Oxford: Blackwell.

Ockwell, D. (2008) 'Energy and Economic Growth: Grounding Our Understanding in Physical Reality', *Energy Policy* 36: 4600–4604.

Oi, J. (1992) 'Fiscal Reform and the Economic Foundations of Local State Corporatism in China', *World Politics* 45(1): 99–126.

Olfman, S. (ed.) (2006) *No Child Left Different*, Westport, CT: Praeger.

Organisation for Economic Cooperation and Development (OECD) (1995) *The Implications of the Knowledge-Based Economy for Future Science and Technology Policies*, OCDE/GD(95) 136, Paris: OECD.

OECD (1996) *Employment and Growth in the Knowledge-Based Economy*, Paris: OECD

OECD (2005) *Proposal for a Major Project on the Bioeconomy in 2030*, Paris: OECD.

OECD (2008) *OECD Reviews of Innovation Policy: China*, Paris: OECD.

Orsi, F. and B. Coriat (2006) 'The New Role and Status of Intellectual Property Rights in Contemporary Capitalism', *Competition & Change* 10(2): 162–179.

Orsenigo, L. (1989) *The Emergence of Biotechnology – Institutions and Markets in Industrial Innovation*, London: Pinter Publishers.

Owen-Smith, J. and W. Powell (2003) 'The Expanding Role of University Patenting in the Life Sciences: Assessing the Importance of Experience and Connectivity', *Research Policy* 32: 1695–1711.

Pagano, U. and M.A. Rossi (2009) 'The Crash of the Knowledge Economy', *Cambridge Journal of Economics* 33: 665–683.

Parsons, S.D. (1999a) 'Why the 'Transcendental' in Transcendental Realism?' in S. Fleetwood (ed.), *Critical Realism in Economics*, London and New York: Routledge.

Parsons, S.D. (1999b) 'Economics and Reality: A Philosophical Critique of Transcendental Realism', *Review of Political Economy* 11(4): 455–466.

Patel, P. and K. Pavitt (1998) 'National Systems of Innovation under Strain: The Internationalisation of Corporate R&D', *SPRU Working Paper*, Falmer: SPRU.

Patomäki, H. (2002) *After International Relations: Critical Realism and the (Re)Construction of World Politics*, London: Routledge.

Pavitt, K. (1998) 'Technologies, Products and Organization in the Innovating Firm: What Adam Smith Tells Us and Joseph Schumpeter Doesn't', *Industrial & Corporate Change* 7(3): 433–452.

Peck, J. and A. Tickell (2002) 'Neoliberalizing Space', *Antipode* 34: 380–404.

Pedersen, M. (2010) 'Property, Commoning and the Politics of Free Software', Special Issue of *The Commoner*, 14.

Perez, C. (2002) *Technological Revolutions and Financial Capital – The Dynamics of Bubbles and Golden Ages*, Cheltenham and Northampton, MA: Edward Elgar.

Pharmaceutical Research and Manufacturers of America (PhRMA) (2005) *Pharmaceutical Industry Profile 2005*, Washington, DC: PhRMA.

Polanyi, K. (1957/1944) *The Great Transformation*, Boston: Beacon Press.

Polanyi, M. (1969) *Knowing and Being*, Chicago: University of Chicago Press.

Porter, T. (1995) *Trust in Numbers: The Pursuit of Objectivity in Science and Public Life*, Princeton, NJ: Princeton University Press.

Poulantzas, N. (1978) *Political Power and Social Classes*, London: Verso.

Powell, W. and K. Snellman (2004) 'The Knowledge Economy', *Annual Review of Sociology* 30: 199–220.

Prevezer, M. (2008) 'Technology Policies in Generating Biotechnology Clusters: A Comparison of China and the US', *European Planning Studies* 16(3): 359–374.

Psillos, S. (1999) *Scientific Realism: How Science Tracks Truth*, London and New York: Routledge.

Public Citizen (2001a) 'The Other Drug War: Big Pharma's 625 Washington Lobbyists', (www.citizen.org).

Public Citizen (2001b) 'Rx R&D Myths: The Case Against the Drug Industry's R&D Scare Card' (www.citizen.org).

Public Citizen (2002) 'The Other Drug War II' (www.citizen.org).

Public Citizen (2003a) 'The Other Drug War 2003' (www.citizen.org).

Public Citizen (2003b) '2002 Drug Industry Profits: Hefty Pharmaceutical Company Margins Dwarf Other Industries' (www.citizen.org).

Public Citizen (2004) 'The Medicare Drug War' (www.citizen.org).

Quine, W.V.O. (1980) *From a Logical Point of View* (2nd edition), Cambridge, MA and London: Harvard University Press.

Radder, H. (2010) 'The Commodification of Academic Research', in H. Radder (ed.), *The Commodification of Academic Research: Science and the Modern University*, Pittsburgh, PA: University of Pittsburgh Press.

Rai, A. (1999) 'Regulating Scientific Research: Intellectual Property Rights and the Norms of Science', *Northwestern University Law Review* 94: 77–152.

Rai, A. and R. Eisenberg (2003) 'Bayh-Dole Reform and the Progress of Biomedicine', *Law and Contemporary Problems* 66: 289–314.

Reddy, P. (2000) *The Globalization of Dorporate R&D: Implications for Host Countries*, London: Routledge.

Rescher, N. (1989) *Cognitive Economy*, Pittsburgh, PA: Pittsburgh University Press.

Richards, D. (2002) 'The Ideology of Intellectual Property Rights in the International Economy', *Review of Social Economy* 60(4): 521–541.

Richards, D. (2004) *Intellectual Property Rights and Global Capitalism – The Political Economy of the TRIPS Agreement*, Armonk, NY and London: M.E. Sharpe.

Rodrigues, M.J. (2003) *European Policies for a Knowledge Economy*, Cheltenham: Edward Elgar.

Ronit, K. (1997) 'Academia-Industry-Government Relations in Biotechnology: Private, Professional and Public Dimensions of the New Associations', *Science and Public Policy* 24(6): 421–433.

Rose, N. (2007a) 'Molecular Biopolitics, Somatic Ethics and the Spirit of Biocapital', *Social Theory and Health* 5: 3–29.

Rose, N. (2007b) *The Politics of Life Itself*, Princeton, NJ: Princeton University Press.

Rosenberg, N. (1974) 'Science, Invention and Economic Growth', *Economic Journal* 84(333): 90–108.

Rosenberg, N. and R. Nelson (1994) 'American Universities and Technical Advance in Industry', *Research Policy* 23: 323–348.

Royal Society (2003) *Keeping Science Open: The Effects of Intellectual Property Policy on the Conduct of Science*, London: The Royal Society.

Royal Society (2009) *Reaping the Benefits: Science and the Sustainable Intensification of Global Agriculture*, London: The Royal Society.

Royal Society (2010) *The Scientific Century: Securing our Future Prosperity*, London: The Royal Society.

Ruccio, D.F. (2005) '(Un)Real Criticism', *Post-Autistic Economics Review* 35: 44–49.

Rudy, A., D. Coppin, J. Konefal, B. Shaw, T. Ten Eyck, C. Harris and L. Busch (2007) *Universities in the Age of Corporate Science: The UC Berkeley-Novartis Controversy*, Philadelphia: Temple University Press.

Sanandaji, N., A. Malm and T. Sanandaji (2010) *The Spirit Illusion*, London: Taxpayers' Alliance.

Saunders, P. (2010) *Beware False Prophets: Equality, The Good Society and The Spirit Level*, London: Policy Exchange.

Savran, S. and E.A. Tonak (1999) 'Productive and Unproductive Labour: An Attempt at Clarification and Classification', *Capital & Class* 68: 113–152.

Sayer, A. (1992) *Method in Social Science: A Realist Approach*, London: Routledge.

Sayer, A. (2000) *Realism and Social Science*, London and Thousand Oaks, CA: Sage.

Schumpeter, J.A. (1976) *Capitalism, Socialism and Democracy* (5th edition), London: George Allen & Unwin.

Schwaag-Serger, S. and M. Breidne (2007) 'China's Fifteen-Year Plan for Science and Technology', *Asia Policy* 4: 135–164.

Scott, J.C. (1998) *Seeing Like a State*, New Haven, CT: Yale University Press.

Segal, A. (2003) *Digital Dragon*, Ithaca, NY: Cornell University Press.

Sell, S. (2003) *Private Power, Public Law: The Globalization of Intellectual Property Rights*, Cambridge: Cambridge University Press.

Sell, S. (2007) 'TRIPs-Plus Free Trade Agreements and Access to Medicines', *Liverpool Law Review* 28: 41–75.

Sell, S. (2010) 'The Rise and Rule of a Trade-Based Strategy: Historical Institutionalism and the International Regulation of Intellectual Property', *Review of International Political Economy* 17(4): 762–790.

Senker, J. (2000) 'Biotechnology: Scientific Progress and Social Progress', in J. de la Mothe and J. Niosi (eds), *Economics and Social Dynamics of Biotechnology*, London: Kluwer Academic Publishers.

Sent, E.-M. (1997) 'An Economist's Glance at Goldman's Economics', *Philosophy of Science* 64 (Proceedings): S139–S148.

Sent, E.-M. (1999) 'Economics of Science: Survey and Suggestions', *Journal of Economic Methodology* 6: 95–124.

Shapin, S. (2008) *The Scientific Life: A Moral History of a Late Modern Vocation*, Chicago: University of Chicago Press.

Shapin, S. and S. Shaffer (1985) *Leviathan and the Air Pump: Hobbes, Boyle, and the Experimental Life*, Princeton, NJ: Princeton University Press.

Shaxson, N. (2011) *Treasure Islands: Tax Havens and the Men Who Stole the World*, London: Bodley Head.

Shiva, V. (1993) *Monocultures of the Mind*, London: Zed Books.

Shiva, V. (2001) *Protect or Plunder: Understanding Intellectual Property Rights*, London and New York: Zed Books.

Shore, C. (2008) 'Audit Culture and Illiberal Governance: Universities and the Politics of Accountability', *Anthropological Theory* 8(3): 278–298.

Sigurdson, J. (2005) *Technological Superpower China*, Cheltenham: Edward Elgar.

Simon, H. (1979) 'Rational Decision Making in Business Organizations', *American Economic Review* 69(4): 493–513.

Sismondo, S. (2004) *An Introduction to Science and Technology Studies*, Oxford: Blackwell.

Sismondo, S. (2007) 'For Realism and Anti-Realism', in J. Frauley and F. Pearce (eds), *Critical Realism and the Social Sciences: Heterodox Elaborations*, Toronto: University of Toronto Press.

Slaughter, S. and S. Leslie (1997) *Academic Capitalism: Politics, Policies and the Entrepreneurial University*, Baltimore and London: Johns Hopkins University Press.

Slaughter, S. and G. Rhoades (2002) 'The Emergence of a Competitiveness Research and Development Policy Coalition and the Commercialization of Academic Science and Technology', in P. Mirowski and E.-M. Sent (eds), *Science Bought and Sold*, Chicago: University of Chicago Press.

Slaughter, S. and G. Rhoades (2004) *Academic Capitalism and the New Economy*, Baltimore and London: John Hopkins University Press.

Smith, K. (2010) 'Research, Policy and Funding – Academic Treadmills and the Squeeze on Intellectual Spaces', *British Journal of Sociology* 61(1): 176–195.

Smith, N. (2007) 'Nature as Accumulation Strategy', *Socialist Register* 43: 16–36.

Snowdon, C. (2010) *The Spirit Level Delusion: Fact-Checking the Left's New Theory of Everything*, London: Democracy Institute.

Soley, L. (1995) *Leasing the Ivory Tower: The Corporate Takeover of Academia*, Boston: South End Press.

Solomon, M. (1995) 'Legend Naturalism and Scientific Progress: An Essay on Philip Kitcher's *The Advancement of Science*', *Studies in History and Philosophy of Science* 26: 205–218.

Steinfeld, E. (2004) 'China's Shallow Integration: Networked Production and the New Challenges for Late Industrialization', *World Development* 32(11): 1971–1987.

Stengers, I. (2010) *Cosmopolitics I* (trans. R. Bononno), Minneapolis, MN: University of Minnesota Press.

Stephens, P. (2010) 'Three Years On, The Markets are Masters Again', *Financial Times* 29 July.

Stern, R. (2003) *Transcendental Arguments: Problems and Prospects*, Oxford: Clarendon Press.

Stiglitz, J. (2003) *Globalization and Its Discontents*, Harmondsworth: Penguin.

Stirling, A. (2001) 'The Precautionary Principle in Science and Technology', in T. O'Riordan and A. Jordan (eds), *Reinterpreting the Precautionary Principle*, London: Cameron-May.

Stirling, A. (2005) 'Opening Up or Closing Down: Analysis, Participation and Power in the Social Appraisal of Technology', in M. Leach, I. Scoones and B. Wynne (eds), *Science, Citizenship and Globalisation*, London: Zed.

Stirling, A. (2009) 'Direction, Distribution and Diversity! Pluralising Progress in Innovation, Sustainability and Development', *STEPS Working Paper 32*, Brighton: STEPS Centre.

Strange, S. (1986) *Casino Capitalism*, Oxford: Blackwell.

Strange, S. (1998) *States and Markets* (2nd edition), London: Pinter.

Strawson, P.F. (1950) 'On Referring', *Mind* 59(235): 320–344.

Strawson, P.F. (1954) 'A Reply to Mr. Sellars', *The Philosophical Review* 63(2): 216–231.

Strawson, P.F. (1956) 'Singular Terms, Ontology and Identity', *Mind* 65(260): 433–454.

Strawson, P.F. (1957) 'Propositions, Concepts and Logical Truths', *The Philosophical Quarterly* 7(26): 15–25.

Strawson, P.F. (1966) *The Bounds of Sense*, London: Methuen.

Stroud, B. (2000) 'The Synthetic A Priori in Strawson's Kantianism', in *Understanding Human Knowledge*, Oxford: Oxford University Press.

Sum, N.-L. (2004) 'Information Capitalism and the Remaking of "Greater China": Strategies of Siliconization', in F. Mengin (ed.), *Cyber China*, New York and Basingstoke: Palgrave Macmillan.

Sun, L. and G. Escobar (1999) 'On Chicken's Front Line', *Washington Post*, 28 November: A1.

Sunder Rajan, K. (2003) 'Genomic Capital: Public Cultures and Market Logics of Corporate Biotechnology', *Science as Culture* 12(1): 87–121.

Sunder Rajan, K. (2006) *Biocapital – The Constitution of Postgenomic Life*, Durham, NC and London: Duke University Press.

Suttmeier, R.P., C. Cao and D.F. Simon (2006) 'China's Innovation Challenge and the Remaking of the CAS', *Innovations*, Summer: 78–97.

Swann, J. (1988) *Academic Scientists and the Pharmaceutical Industry: Cooperative Research in Twentieth Century America*, Baltimore: Johns Hopkins University Press.

Tansey, G. and T. Worsley (1995) *The Food System*, London: Earthscan.

Tapscott, D. and A. Williams (2008) *Wikinomics*, New York: Atlantic Books.

Taylor C. and Z. Silberstom (1973) *The Economic Impact of the Patent System*, Cambridge: Cambridge University Press.

Thackray, A. (ed.) (1998) *Private Science: Biotechnology and the Rise of the Molecular Sciences*, Philadelphia: University of Pennsylvania Press.

Thompson, E.P. (1967) *The Making of the English Working Class*, New York: Vintage Books.

Thrift, N. (2005) *Knowing Capitalism*, London: Sage.

Thursby, J. and M. Thursby (2003) 'University Licensing and the Bayh-Dole Act', *Science* 301: 1052.

Tickell, A. and J. Peck (2003) 'Making Global Rules: Globalization or Neoliberalization?' in J. Peck and H. Yeung (eds), *Remaking the Global Economy*, London: Sage.

Tijssen, R. (2004) 'Is the Commercialisation of Scientific Research Affecting the Production of Public Knowledge? Global Trends in the Output of Corporate Research Articles', *Research Policy* 33: 709–733.

Tyfield, D. (2007) 'Chasing Fairies or Serious Ontological Business: Tracking Down the Transcendental Argument', in C. Lawson, J. Latsis and N. Martin (eds), *Contributions to Social Ontology*, London and New York: Routledge.

Tyfield, D. (2008a) 'Enabling TRIPs: The Pharma-Biotech-University Patent Coalition', *Review of International Political Economy* 15(4): 535–566.

Tyfield, D. (2008b) 'The Impossibility of Finitism: From SSK to ESK?', *Erasmus Journal of Philosophy and Economics* 1(1): 61–86.

Tyfield, D. (2008c) 'Raging at Imaginary Don Quixotes: Reply to Giraud and Weintraub', *Erasmus Journal of Philosophy and Economics* 2(1): 60–69.

Tyfield, D. (2009) 'Review: A Surplus of 'Surplus', *Science as Culture* 18(4): 497–500.

Underhill, G. (2006) 'Introduction: Conceptualizing the Changing Global Order', in R. Stubbs and G. Underhill (eds), *Political Economy and the Changing Global Order* (3rd edition), Oxford and New York: Oxford University Press.

United States Patent and Trademarks Office (USPTO) (2010) 'Calendar Year Patent Statistics', Commissioner of Patents and Trademarks Annual Report, Washington, DC: USPTO, available at www.uspto.gov/web/offices/ac/ido/oeip/taf/reports.htm.

Urry, J. (1995) *Consuming Places*, London and New York: Routledge.

van den Belt, H. (2010) 'Robert Merton, Intellectual Property, and Open Science: A Sociological History for Our Times', in H. Radder (ed.), *The Commodification of Academic Research: Science and the Modern University*, Pittsburgh, PA: University of Pittsburgh Press.

van der Ploeg, J.D. (2010) 'The Food Crisis, Industrialized Farming and the Imperial Regime', *Journal of Agrarian Change* 10(1): 98–106.

Vandenberghe, F. (2002) 'Reconstructing Humants: A Humanist Critique of Actant-Network Theory', *Theory, Culture & Society* 19(5/6): 51–67.

Vercellone, C. (2007) 'From Formal Subsumption to *General Intellect*: Elements for a Marxist Reading of the Thesis of Cognitive Capitalism', *Historical Materialism* 15(1): 13–36.

Viskovatoff, A. (2002) 'Critical Realism and Kantian Transcendental Arguments', *Cambridge Journal of Economics* 26: 697–708.

von Hippel, E. (2005) *Democratizing Innovation*, Cambridge, MA: MIT Press.

Vromen, J. (2004) 'Conjectural Revisionary Ontology', *Post-Autistic Economics Review* 29: 16–22.

Wade, R. (2006) 'Choking the South', *New Left Review* 38: 115–127.

Waldfogel, J. (1998) 'Reconciling Asymmetric Information and Divergent Expectations Theories of Litigation', *Journal of Law and Economics*, October.

Walsh, W.H. (1975) *Kant's Criticism of Metaphysics*, Edinburgh: Edinburgh University Press.

Wang, J.-H. (2006) 'China's Dualist Model on Technological Catching Up: A Comparative Perspective', *The Pacific Review* 19(3): 385–403.

Wang, B. and J. Ma (2007) 'Collaborative R&D: Intellectual Property Rights between Tsinghua University and Multinational Companies', *Journal of Technology Transfer* 32: 457–474.

Wang, Y. and S. Johnston (2007) 'Review on GM Rice Risk Assessment in China', *UNU-IAS Working Paper No. 152*, April.

Ward K. (2005) 'Geography and Public Policy: A Recent History of "Policy Relevance"', *Progress in Human Geography* 29(3): 310–319.

Washburn, J. (2005) *University Inc.: The Corporate Corruption of American Higher Education*, New York: Basic Books.

Webster, A. (1994) 'University-Corporate Ties and the Construction of Research Agendas', *Sociology* 28: 123–142.

Weis, T. (2007) *The Global Food Economy: The Battle for the Future of Food and Farming*, London: Zed Books.

Welsh, R. and L. Glenna (2006) 'Considering the Role of the University in Conducting Research on Agri-Biotechnologies', *Social Studies of Science* 36(6): 929–942.

Whatmore, S. and L. Thorne (1998) 'Reconfiguring the Geographies of Wildlife', *Transactions of the Institute of British Geographers* 23: 435–454.

Wible, J. (1998) *The Economics of Science*, London: Routledge.

Wilkinson, R. and K. Pickett (2009) *The Spirit Level: Why Equality is Better for Everyone*, London: Penguin.

Williams, O. (2000) 'Life Patents, TRIPs and the International Political Economy of Biotechnology', in J. Vogler and A. Russell (eds), *The International Politics of Biotechnology*, Manchester and New York: Manchester University Press.

Wilsdon, J. and J. Keeley (2007) *China: The Next Science Superpower*, London: Demos.

Winickoff, D., S. Jasanoff, L. Busch, R. Grove-White and B. Wynne (2005) 'Adjudicating the GM Food Wars: Science, Risk and Democracy in World Trade Law', *Yale Journal of International Law* 30: 81–124.

Wood, E.M. (2000) 'The Agrarian Origins of Capitalism', in F. Buttel, H. Magdoff and J.B. Foster (eds), *Hungry for Profit: The Agribusiness Threat to Farmers, Food and the Environment*, New York: Monthly Review Press.

World Bank (2010) 'Royalties and Licence Fees', Washington, DC: World Bank, available at http://data.worldbank.org/indicator/BX.GSR.ROYL.CD.

Wright, S. (1994) *Molecular Politics: Developing American and British Regulatory Policy for Genetic Engineering, 1972–1982*, Chicago: University of Chicago Press.

Wright, S. (1998) 'Molecular Politics in a Global Economy', in A. Thackray (ed.), *Private Science: Biotechnology and the Rise of the Molecular Sciences*, Philadelphia: University of Pennsylvania Press.

Wu, W. (2007) 'Cultivating Research Universities and Industrial Linkages in China', *World Development* 35(6): 1075–1093.

Wynne, B. and U. Felt (2007) *Taking European Knowledge Society Seriously*, Science Economy and Society Directorate EUR 22700, Brussels: European Commission D-G Research.

Xin, H. (2008) 'You Say You Want a Revolution', *Science*, 31 October, 322: 664–666.

Xue L. and N. Forbes (2006) 'Will China Become a Science and Technology Superpower by 2020?' *Innovations* (Fall): 111–126.

Zamorra Bonilla, J. (1999) 'Elementary Economics of Scientific Consensus', *Theoria* 14: 461–488.

Zeller, C. (2008) 'From the Gene to the Globe', *Review of International Political Economy* 15(1): 86–115.

Zhao, J.H. and P. Ho (2005) 'A Developmental Risk Society? The Politics of Genetically Modified Organisms (GMOs) in China', *International Journal of Environment and Sustainable Development* 4(4): 370–394.

Zhou, Y. (2005) 'The Making of an Innovative Region from a Centrally Planned Economy', *Environment and Planning A* 37: 1113–1134.

Zhou, P. and L. Leydesdorff (2006) 'The Emergence of China as a Leading Nation in Science', *Research Policy* 35: 83–104.

Zhou, X., W. Zhao, Q. Li and H. Cai (2003) 'Embeddedness and Contractual Relationships in China's Transitional Economy', *American Sociological Review* 68: 75–102.

Zhu, D. and J. Tann (2005) 'A Regional Innovation System in a Small-sized Region', *Technology Analysis & Strategic Management* 17(3): 375–390.

INDEX

Note: page numbers in *italics* refer to figures, those in **bold** refer to tables, and those followed by n refer to the notes at the end of the book.

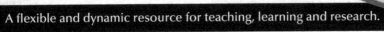